T0205757

# Lecture Notes in Computer Science    11413

*Commenced Publication in 1973*
Founding and Former Series Editors:
Gerhard Goos, Juris Hartmanis, and Jan van Leeuwen

More information about this series at http://www.springer.com/series/7407

Sebastian Feld · Claudia Linnhoff-Popien (Eds.)

# Quantum Technology and Optimization Problems

First International Workshop, QTOP 2019
Munich, Germany, March 18, 2019
Proceedings

 Springer

*Editors*
Sebastian Feld
Ludwig-Maximilians-Universität München
Munich, Germany

Claudia Linnhoff-Popien
Ludwig-Maximilians-Universität München
Munich, Germany

ISSN 0302-9743        ISSN 1611-3349  (electronic)
Lecture Notes in Computer Science
ISBN 978-3-030-14081-6        ISBN 978-3-030-14082-3  (eBook)
https://doi.org/10.1007/978-3-030-14082-3

Library of Congress Control Number: 2019931877

LNCS Sublibrary: SL1 – Theoretical Computer Science and General Issues

This Springer imprint is published by the registered company Springer Nature Switzerland AG
The registered company address is: Gewerbestrasse 11, 6330 Cham, Switzerland

# Preface

Over the past decade, a wide variety of experimental quantum computing hardware has been invented and used for fundamental demonstrations in laboratories. Initial results confirm the feasibility of such hardware in real-world applications. Recently, one can observe upcoming research in areas like traffic flow optimization, mobile sensor placement, machine learning, and many more. The development of quantum computing hardware, be it in the quantum gate model or adiabatic quantum computation (quantum annealing), has made huge progress in the past few years. This started to transfer know-how from quantum technology-based research to algorithms and applications. This development is offering numerous opportunities to contribute within research, theory, applied technologies, and engineering.

The First International Workshop on Quantum Technology and Optimization Problems (QTOP 2019) was held in conjunction with the International Conference on Networked Systems (NetSys 2019) in Munich, Germany, on March 18, 2019.

The aim of this workshop was to connect people from academia and industry to discuss theory, technology, and applications and to exchange ideas in order to move efficiently forward in engineering and development in the exciting area of quantum technology and optimization problems.

This book comprises a section containing a keynote and four sections with scientific papers. The sessions deal with the following topics that are crucial to the development of future improvements in quantum technology and optimization problems:

- Analysis of optimization problems
- Quantum gate algorithms
- Applications of quantum annealing
- Foundations of quantum technology

The international call for papers resulted in the selection of the papers in these proceedings. The selection of the papers followed a rigorous review process involving an international expert group. Each paper was reviewed by at least three reviewers.

We express our gratitude to all members of the Program Committee for their valuable work. We also want to thank the members of the Mobile and Distributed Systems Group, and especially the Quantum Applications and Research Laboratory (QAR-Lab) who were responsible for the organization of the conference.

March 2019

Sebastian Feld
Claudia Linnhoff-Popien

# Organization

## General Chairs

| | |
|---|---|
| Sebastian Feld | LMU Munich, Germany |
| Claudia Linnhoff-Popien | LMU Munich, Germany |

## Program Committee

| | |
|---|---|
| Nick Chancellor | Durham University, UK |
| Bo Ewald | D-Wave Systems, Canada |
| Markus Friedrich | LMU Munich, Germany |
| Thomas Gabor | LMU Munich, Germany |
| Markus Hoffmann | Google, Germany |
| Faisal Shah Khan | Khalifa University, Abu Dhabi |
| Dieter Kranzlmüller | LRZ, Germany |
| Luke Mason | Science & Technology Facilities Council, UK |
| Wolfgang Mauerer | OTH Regensburg, Germany |
| Catherine C. McGeoch | D-Wave Systems, Canada |
| Masayuki Ohzeki | Tohoku University, Japan |
| Jonathan Olson | Zapata Quantum Computing, USA |
| Dan O'Malley | Los Alamos National Laboratory, USA |
| Tobias Stollenwerk | DLR, Germany |

## Organizing Committee

| | |
|---|---|
| Carsten Hahn | LMU Munich, Germany |
| Stefan Langer | LMU Munich, Germany |
| Robert Müller | LMU Munich, Germany |
| Christoph Roch | LMU Munich, Germany |
| Andreas Sedlmeier | LMU Munich, Germany |

# Contents

**Foundations and Quantum Technologies**

# Keynote

# An Introduction to Quantum Computing and Its Application

Robert H. (Bo) Ewald[(⊠)]

D-Wave Systems, Burnaby, Canada
bewald@dwavesys.com

**Abstract.** This paper reviews the two major approaches to quantum computing and discusses early quantum applications.

**Keywords:** Quantum computing · Quantum applications

## 1 Getting Started

Traditional (or classical) digital computers are running into barriers that limit continued performance improvement as many have written about with regard to the end of Moore's Law. But, some aspects of the fundamental physics that create challenges as we seek to cram more transistors onto a chip may also enable a different form of computing. That new type of computing is quantum computing and even though the idea has been around for 35+ years, we are just now arriving at the "end of the beginning" to borrow from Winston Churchill.

Nobel prize-winning physicist Richard Feynman proposed in 1981 that perhaps a quantum mechanical computer could be created that would better model or simulate nature and let us understand quantum behavior and solve previously intractable problems. The core of his idea was to create a quantum computer, program it, let it run, and it would behave in a quantum manner as nature does. Quantum logic says this should work better than using a classical computer to model quantum reality. Researchers around the world continued working on quantum theory and began developing quantum devices, computers, communications, and the technology and architectures to create them.

## 2 Quantum Computing Architectures and Technology

When we design and build a computer, digital or quantum, we match and marry two things – the system architecture and the technology. The system architecture lays out how the computer works and we match that architecture to a technology that we use to implement the system. The architecture specifies a set of instructions that cause the computer to perform certain actions.

© Springer Nature Switzerland AG 2019
S. Feld and C. Linnhoff-Popien (Eds.): QTOP 2019, LNCS 11413, pp. 3–8, 2019.
https://doi.org/10.1007/978-3-030-14082-3_1

In a digital computer one of the fundamental building blocks is a "bit" – one piece of information that is stored as zero or a one. A string of digital bits can represent a number, a letter, or a sentence for example. Once the value of a digital bit is set, it doesn't change until the computer causes it to change.

In a quantum computer, the basic building block is a quantum bit, or a "qubit." If we get our qubit into a state of quantum superposition, the qubit is a zero and a one simultaneously. We can now routinely build qubits from superconducting circuits (D-Wave, IBM, Google, Rigetti, and others have done this). Qubits can also be created by "trapping" ions with lasers (IonQ). Microsoft hopes to create a "topological" qubit from a quasi-particle, using superconducting technology. However, all qubits are not created equal. Depending on the technology chosen, some operate in wholly different ways than others, some have more errors, some have longer coherence times, and some are faster than others [1].

## Gate Model Architecture

The first quantum system architecture that was investigated starting in the 1980's was the "gate" or "circuit" model architecture. That followed naturally from digital computers in which a designer has connected a series of digital logic gates ("AND", "OR", "NOT", ...) to perform some function like adding two numbers. But, in the quantum gate model systems, the qubits aren't simple digital zeros or ones. They can be simultaneously a zero and a one, and relate to each other, so we have a more complicated set of gates to operate with (Hadamard, Pauli gates, Toffoli, etc.). The application connects and strings together the gates and qubits upon which they operate as shown in the example in Fig. 1 from the Jülich Supercomputer Center with a 5 qubit IBM system performing a simple unit test [2].

One of the challenges with the gate model architectures has to do with qubit quality and/or error correction. The gate model system needs to stay in a coherent quantum state for the entirety of the calculation, or the results me be suspect or there may be no result. There have been several analyses over the past 20 years of the level of quality that qubits need to achieve, or the amount of error correcting qubits that would be required which for large general purpose gate model systems may be in the millions. However, the industry is many years if not decades from achieving systems with millions of qubits [3, 4]. Alternatively, smaller gate model systems might be able to perform some useful calculations with the current qubit quality and no error correcting qubits. Systems like this are being called Noisy Intermediate Scale Quantum Computers (NISQ) [5] or Approximate Quantum Computers. The next year should determine if this idea is useful and for which application areas. And, some technologies promise much higher qubit quality, and people will soon be able to work on those systems.

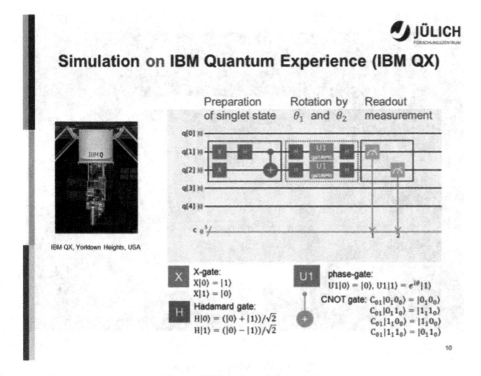

**Fig. 1.** IBM gate model example

**Quantum Annealing Architecture**

About 20 years ago, Prof. Hidetoshi Nishimori and his PhD student Tadashi Kadowaki at the Tokyo Institute of Technology first proposed a second quantum computer architecture that today is called "quantum annealing." [6] The idea behind quantum annealing is quite different than the gate model design, and is more like an analog computer architecture. The basic idea is that a wide range of important problems can be represented as a mountainous landscape (imagine the Alps), and a quantum annealing computer will find the low valley in the landscape, probably. Of course, it isn't really a three-dimensional mountainous landscape, but it is a high-dimensional energy landscape and rather than a valley, we are finding a low-energy solution. Annealing quantum computers are designed to find good (not necessarily perfect) solutions to this problem, very quickly. Whereas a classical computer would have to visit and test each valley, the quantum annealer can visit and test many valleys simultaneously—due to the operation of the quantum effects present in the system. Figure 2 shows the D-Wave 2000 qubit processor, the actual computer is a single chip.

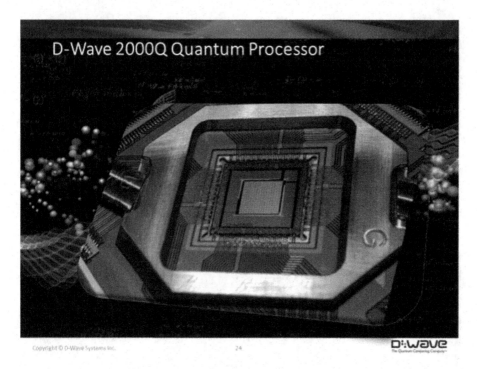

**Fig. 2.** D-Wave 2000 qubit processor

## 3   Early Quantum Applications

D-Wave's quantum annealing systems were the first commercially available quantum systems in the market. Lockheed Martin and USC's Information Sciences Institute installed the first D-Wave system in 2011; Google and their partners, NASA Ames Research Center and USRA, installed a 500 qubit D-Wave system about five years ago and have upgraded twice, including installing the first 2000 qubit D-Wave 2000Q™ system in the fall of 2017; and Los Alamos National Laboratory became D-Wave's third system customer, installing a 1000 qubit system in 2016 which has just been upgraded to 2000 qubits. About 40 other organizations including Oak Ridge National Laboratory, Volkswagen, Toyota Tsusho, DENSO, Recruit Communications, Nomura Securities and several universities have contracts with D-Wave and are working on 2000 qubit systems over the cloud.

These customers have developed about 100 prototype applications (proto-apps), about half of which are in optimization, 20% in machine learning, 10% in material science and a wide range of other areas. Some of these proto-apps have shown that the D-Wave systems are approaching and sometimes surpassing conventional computing in terms of performance or solution quality, heralding the first examples of real customer application advantage on quantum computers.

However, the systems are not yet large enough (in terms of qubit count and connections between qubits) to be able to run production size problems. As an example, in highly publicized work in 2017, Volkswagen used a 1000 qubit D-Wave system to optimize routing of taxis in Beijing. Using the quantum hardware and classical computer partitioning software, the system found better routes for about 500 taxis from downtown Beijing to the airport to minimize overall congestion. But the larger routing problem of the roughly 10,000 actual taxis in Beijing was too large for today's systems. As the systems continue to grow in qubits, connectivity and other functions, and as software partitioning tools improve, production size applications will be in reach. Details on many of the proto-apps are at the following websites that have collected presentations from the first four D-Wave Qubits Users Group Meetings, Los Alamos' first 22 "Rapid Response Projects" and videos from Denso about their applications.

D-Wave Users Group Presentations:

- https://dwavefederal.com/qubits-2016/
- https://dwavefederal.com/qubits-2017/
- https://www.dwavesys.com/qubits-europe-2018
- https://www.dwavesys.com/qubits-north-america-2018

LANL Rapid Response Projects:

- http://www.lanl.gov/projects//national-security-education-center/information-science-technology/dwave/index.php

DENSO Videos:

- https://www.youtube.com/watch?v=Bx9GLH_GklA (CES – Bangkok)
- https://www.youtube.com/watch?v=BkowVxTn6EU (CES – Factory)
- https://www.youtube.com/watch?v=4zW3_lhRYDc (AGV's).

## 4 The "End of the Beginning"

It looks like the next year or two (2019–2020) will prove to be a time when the applications will become more clear for both types of quantum computing architectures. With the gate model architectures, we expect to see if some applications can run on these NISQ or Approximate machines without built-in error correction. Many people believe that quantum material science/chemistry algorithms may be good candidates. If so, then that will chart one direction. If not, it is probably back to the drawing board to look at other candidate applications while error correction, qubit quality, and/or ion trap or other more fault resistant technologies and architectures are developed.

For the quantum annealing architecture, the path is perhaps more clear. D-Wave has been able to double the number of qubits in its systems roughly every two years, and the next generation is planned to have 4000–5000 qubits with a big improvement in the connectivity between the qubits and in qubit quality. At that scale, and with improvements in software partitioning to break large production problems into smaller pieces that fit on the hardware, the next generation system will probably be the turning point to scale up quantum computing for real-world problems. Then, in addition to the optimization and machine learning proto-apps on D-Wave systems, we should start to see more quantum material science applications. Those applications should open new ways of modeling quantum material behavior and start fulfilling the dream of Richard Feynman. In terms of performance, we should see more customer proto-apps that approach and surpass conventional computer performance or quality of the results, and perhaps see the first demonstrations of quantum advantage in some areas.

If there is a long delay in the availability of larger gate model systems, other approaches will certainly be explored. For example, Atos in Europe recently introduced the Atos Quantum Learning Machine which is a digital system, but simulates up to 40-qubit gate model systems in its digital hardware, and Fujitsu has demonstrated a digital annealing system.

So, as we reach the end of the beginning era in quantum computing, the way ahead is still indeterminate which is appropriate for things quantum, but it is also full of hope and promise.

# References

1. Popkin, G.: Quest for qubits. Sci. Mag. **354**(6316), 1090–1093 (2016)
2. Michielsen, K., Nocon, M., Willsch, D., Jin, F., Lippert, T., De Raedt, H.: Benchmarking gate-based quantum computers. Julich Supercomputer Centre, Aachen University and the University of Groningen (2017). arXiv:1706.04341v1
3. O'Gorman, J., Campbell, E.T.: Quantum computation with realistic magic-state factories. Phys. Rev. A **95**, 032338 (2017)
4. Suchara, M., Faruque, A., Lai, C.-Y., Paz, G., Chong, F.T., Kubiatowicz, J.: Comparing the overhead of topological and concatenated quantum error correction. arXiv:1312.2316v1
5. Preskill, J.: Quantum Computing in the NISQ era and beyond (2018). arXiv:1801.00862
6. Kadowaki, T., Nishimori, H.: Quantum annealing in the transverse Ising model. Phys. Rev. E **58**(5), 5355 (1988)

# Analysis of Optimization Problems

# Embedding Inequality Constraints for Quantum Annealing Optimization

Tomáš Vyskočil[2], Scott Pakin[1], and Hristo N. Djidjev[1(✉)]

[1] Los Alamos National Laboratory, Los Alamos, New Mexico, USA
{pakin,djidjev}@lanl.gov
[2] Rutgers University, New Brunswick, New Jersey, USA
xwhisky@gmail.com

**Abstract.** Quantum annealing is a model for quantum computing that is aimed at solving hard optimization problems by representing them as quadratic unconstrained binary optimization (QUBO) problems. Although many NP-hard problems can easily be formulated as binary-variable problems with a quadratic objective function, such formulations typically also include constraints, which are not allowed in a QUBO. Hence, such constraints are usually incorporated in the objective function as additive penalty terms. While there is substantial previous work on implementing linear equality constraints, the case of inequality constraints has not much been studied. In this paper, we propose a new approach for formulating and embedding inequality constraints as penalties and describe early implementation results.

## 1 Introduction

*Quantum annealing* is a hardware analogue of simulated annealing [11], a well-known black-box approach for solving optimization problems. (By "black-box" we mean that no assumptions, e.g., differentiability, can be made of the function to optimize.) However, quantum annealing takes advantage of quantum effects—in particular *quantum tunneling*—to converge with higher probability than simulated annealing given the same annealing schedule [10]. Quantum annealers have been designed to look for solutions of a specific type of binary optimization problem as discussed next.

### 1.1 Quadratic Unconstrained Binary Optimization

Current quantum annealers, such as the D-Wave 2000Q [8], minimize only quadratic pseudo-Boolean functions. That is, they heuristically solve a *quadratic unconstrained binary optimization* problem (QUBO),

$$\arg \min_{x} \overbrace{\left( \sum_{i=1}^{n} a_i x_i + \sum_{i=1}^{n-1} \sum_{j=i+1}^{n} b_{i,j} x_i x_j \right)}^{\mathrm{Obj}(\boldsymbol{x})} \tag{1}$$

© Springer Nature Switzerland AG 2019
S. Feld and C. Linnhoff-Popien (Eds.): QTOP 2019, LNCS 11413, pp. 11–22, 2019.
https://doi.org/10.1007/978-3-030-14082-3_2

for $x \in \mathbb{B}^n$ (where $\mathbb{B}$ indicates the set $\{0,1\}$) given $a \in \mathbb{R}^n$ and $b \in \mathbb{R}^{n \times n}$. We write "heuristically solve" because, just like simulated annealing, global optimality is not guaranteed; a quantum annealer takes only a best-effort approach. Finding the true global minimum is an NP-hard problem [1] so any hardware that can accelerate solutions—even if only by a polynomial amount—can be useful in practice for reducing time to solution.

Although Eq. (1) may seem restrictive in expressiveness, there are in fact a large set of problems with a known mapping into that form [13]. A particular challenge in expressing an optimization problem as a QUBO is that a QUBO by definition does not allow constraints. However, a constraint can be expressed in terms of *penalties*, $a_i x_i$ and $b_{i,j} x_i x_j$ terms that can be included in Eq. (1) and that evaluate to zero when the constraint is honored and to a positive number when the constraint is violated. For example, one can constrain $x_1 = x_2$ by adding a penalty term of $x_1 + x_2 - 2x_1 x_2$ to the QUBO. In this case, $\text{Obj}(0,0) = \text{Obj}(1,1) = 0$, but $\text{Obj}(0,1) = \text{Obj}(1,0) = 1$. A quantum annealer would therefore favor the $x_1 = x_2$ cases when computing $\arg\min_x \text{Obj}(x)$. Kochenberger et al. list a handful of other such penalties in their survey of QUBO problems [12].

## 1.2 Physical Limitations

Equation (1) describes a *logical* optimization problem. In practice, quantum-annealing hardware imposes a number of additional limitations on what can be expressed. Some of the key differences between the logical and physical problems are as follows:

- The number of variables ($n$) is limited to the number of qubits (quantum bits) provided by the quantum processing unit (QPU). In the case of a D-Wave 2000Q, $n \le 2048$.
- The $a$ and $b$ coefficients have neither infinite range nor infinite precision. In the case of a D-Wave 2000Q, each $a_i \in [-2.0, 2.0]$, and each $b_{i,j} \in [-1.0, 1.0]$, with approximately 5–6 bits of precision.
- Only a small subset of the $b_{i,j}$ coefficients are allowed to be nonzero. In the case of a D-Wave 2000Q, there can be at most six nonzero $b_{i,j}$ coefficients for a given $i$ due to the hardware's physical topology.

To elaborate on that final difference, a D-Wave 2000Q employs a particular physical topology called a *Chimera graph* [4]. A nonzero $a_i$ coefficient can appear at any vertex in the graph, but a nonzero $b_{i,j}$ coefficient can appear only where an edge is present in the graph. Figure 1 illustrates a Chimera graph. A Chimera graph is composed of complete, 8-vertex bipartite graphs (i.e., $K_{4,4}$) called *unit cells* (Fig. 1(a)). Each vertex in the first partition connects to its peer to the north and its peer to the south. Each vertex in the second partition connects to its peer to the east and its peer to the west. Consequently, a Chimera graph is, excluding vertices on the boundaries, a degree-six graph, which is quite sparse.

Due to manufacturing and calibration imperfections, some vertices and edges are inevitably absent from the hardware's Chimera graph. For instance, Fig. 1(b)

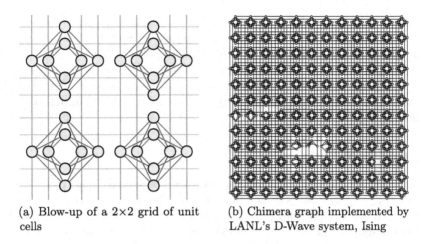

(a) Blow-up of a 2×2 grid of unit cells

(b) Chimera graph implemented by LANL's D-Wave system, Ising

**Fig. 1.** Illustration of the Chimera-graph topology

presents the physical topology of Ising, a D-Wave system installed at Los Alamos National Laboratory (LANL). Ising is missing 11 vertices and 62 edges relative to a complete C12 Chimera graph. (*C12* denotes a $12 \times 12$ grid of unit cells.)

## 1.3   Problem Embedding

We revisit the notion of penalty terms introduced in Sect. 1.1 in light of the hardware limitations described in Sect. 1.2. Consider implementing as a QUBO the constraint $(\sum_{i=1}^{n} x_i) = k$ for some $0 \leq k \leq n$ and with $x_i \in \mathbb{B}$. One would typically accomplish this by adding a "squared error" expression,

$$M \left( (\textstyle\sum_{i=1}^{n} x_i) - k \right)^2 \tag{2}$$

to the quadratic pseudo-Boolean function to be minimized, for a sufficiently large $M$. For example, constraining exactly two of three Boolean variables to be 1 implies setting $n = 3$ and $k = 2$ and deriving $\mathrm{Obj}(\boldsymbol{x})$ as

$$
\begin{aligned}
M \left( (\textstyle\sum_{i=1}^{3} x_i) - 2 \right)^2 &= M(x_1^2 + 2x_1x_2 + 2x_1x_3 - 4x_1 \\
&\quad + x_2^2 + 2x_2x_3 - 4x_2 + x_3^2 - 4x_3 + 4) \\
&= M(-3x_1 - 3x_2 - 3x_3 + 2x_1x_2 + 2x_1x_3 + 2x_2x_3 + 4),
\end{aligned}
$$

recalling that $x_i^2 = x_i$ when $x_i \in \mathbb{B}$. Furthermore, the constant 4 can be removed from the penalty because constant terms do not affect the solution to Eq. (1).

While the "squared error" approach is generally applicable, it suffers from a severe shortcoming in practice: For $n$ variables, it requires $n^2$ quadratic terms. In a graph representation, this implies all-to-all connectivity—a feature lacking in contemporary hardware. A workaround is to *minor-embed* [5] the problem graph into the hardware graph. This implies replacing individual variables with multiple variables and *chaining* them together—constraining them to have the same value using the $x_1 = x_2$ approach presented in Sect. 1.1. For a complete graph, minor embedding incurs a substantial (quadratic) cost in the variable count. Specifically, a $C_m$ Chimera graph can represent problems of no more than $4m + 1$ variables if all $(4m + 1)^2$ quadratic terms are nonzero [3]. For example, even with 1141 vertices, the C12 subgraph shown in Fig. 1(b) is limited to $n = 49$ when applying the Eq. (2) constraint.

### 1.4  Our Contribution

In this paper we introduce a novel approach to represent inequality constraints of the form $\sum_{i=1}^{n} x_i \leq k$ using roughly $kn$ variables. Our approach is based on a set of "gadgets" that map directly to a unit cell and that spatially tile the Chimera graph in a manner that enforces the desired constraint.

The rest of the paper is structured as follows. In Sect. 2 we discuss related efforts to map constraints to QUBOs. Section 3 explains the basics of our approach. The methods used to implement our approach are detailed in Sect. 4. We present in Sect. 5 some empirical analysis of the efficacy of our techniques on actual quantum-annealing hardware. Finally, we draw some conclusions from our work in Sect. 6.

## 2  Related Work

Much work has been done in mapping individual computational problems to QUBOs. Kochenberger et al. survey a set of these dating back to the early 1970s [12]. (Note that Kochenberger et al. use the term *unconstrained binary quadratic programming* or *UBQP* instead of QUBO. The two are synonymous.) Baharona [1] describes how a large set of problems can be mapped to the form of an Ising-model Hamiltonian function, which is structurally identical to Eq. (1), but solves for $x \in \{-1, +1\}^N$ rather than $x \in \{0, 1\}^N$. A simple linear transformation converts between the two.

Perhaps the most closely related work to what we are proposing is Bian et al.'s investigation into mapping constraint satisfaction problems (CSPs) to a D-Wave system while carefully considering the embedding on the Chimera graph [2]. One difference is our focus on scalability that allows constraints of any number of variables to be embedded and that we are able to handle inequalities. We have also applied an approach similar to the one described in Sects. 3 and 4 to equality constraints of the type $\sum_{i=1}^{n} x_i = 1$ [15] and later generalized this approach [14].

## 3  Optimization Based Approach

In this section we define the problem we intend to solve and present an outline of our approach.

## 3.1  Formulating a Constraint Embedding as an Optimization Problem

We first overview the properties of the standard penalty-based approach and then the approach adopted in this paper that uses optimization and extra binary variables.

**Implementation of a Constraint as a Penalty.** Although most NP-hard optimization problems admit simple formulations as optimization problems with quadratic binary objective functions, such formulations typically contain at least one constraint and are therefore not in the QUBO form represented by Eq. (1). In order to convert optimization problems containing constraints into QUBOs, each constraint is usually converted to a penalty that is included in the objective function as an additive term. We call $x = \{x_1, \ldots, x_n\}$ *feasible* if it satisfies the constraint, and *infeasible*, otherwise.

Given a minimization problem with objective $\text{Obj}(x)$ and an equality constraint $C(x) = 0$, the penalty method transforms the constraint into the quadratic term $Q(x) = (C(x))^2$ and changes the objective to $\text{Obj}'(x) = \text{Obj}(x) + MQ(x)$ for a large positive constant $M$. When $x$ is feasible, i.e., if $C(x) = 0$, then $Q(x) = 0$ and the penalty term does not change the value of the objective, i.e., $\text{Obj}'(x) = \text{Obj}(x)$. But if $x$ is infeasible, then $Q(x) \geq \gamma$ and $\text{Obj}'(x) \geq \text{Obj}(x) + M\gamma$, where $\gamma$ is the minimum nonzero value of $Q(x)$. Hence, for the penalty method, the following properties of the penalty function are essential:

(i) If $x$ is feasible, then $Q(x) = 0$;
(ii) if $x$ is infeasible, then $Q(x) \geq \gamma$,

and the parameter $\gamma$, called *gap*, should be as large as possible. Having a large gap is important because it enables using smaller $M$ values, making the resulting optimization problem more stable numerically and thereby increasing the accuracy of the solution.

As discussed in the introduction, implementing a linear constraint as a penalty comes with significant drawbacks. But implementing a linear *inequality* constraint such as

$$\sum_{i=1}^{n} x_i \leq k \tag{3}$$

is even harder because the squared-error approach does not apply. One way to deal with such inequalities is to introduce a new variable $z \geq 0$ such that $(\sum_{i=1}^{n} x_i) + z = k$ and then to apply to the resulting equality constraint the penalty method described above. However, $z$ does not necessarily take binary values in a feasible solution and can be as large as $k$, so it needs to be represented in a binary form by introducing $\lceil \log k \rceil$ binary variables with possibly large ($\approx k$) QUBO coefficients, further complicating the resulting QUBO. Hence, we are looking in this paper at the problem of finding a scalable and qubit-efficient way for embedding inequality-type constraints in quantum annealers such as those by D-Wave Systems.

**Using Ancillary Variables.** One approach, previously examined by multiple researchers [2,15], is to include $m$ ancillary variables $t_i$ and use the additional degrees of freedom to establish a larger "gap" between feasible and infeasible solutions and/or reduce the number of quadratic coefficients needed. Formally, let $t = \{t_1, \ldots, t_m\}$. We want to define a QUBO $Q(x, t)$ that satisfies the following analogues of properties (i) and (ii) above:

**(i′)** If $x$ is feasible, then $\min_t Q(x, t) = 0$;
**(ii′)** if $x$ is infeasible, then $\min_t Q(x, t) \geq \gamma$,

and that also maximizes $\gamma$.

Although these conditions look very similar, (ii′) can be implemented as a linear programming constraint with respect to the real variables $a_i$ and $b_{i,j}$ from Eq. (1), while (i′) is not linear and implementing it requires, e.g., the introduction of additional binary variables, resulting in a much more difficult problem. Moreover, the total number of constraints is $2^{n+m}$ as there is a constraint for each $x \in \mathbb{B}^n$ and each $t \in \mathbb{B}^m$. The resulting mixed-integer programming (MIP) problem with exponential number of constraints cannot be solved in reasonable time for more than 3–4 unit cells of the Chimera graph (i.e., $n + m$ is no more than 30–40). Clearly, this approach cannot be applied to current D-Wave systems with up to 2048 qubits and especially not to future generations with much higher numbers of qubits.

### 3.2 Our Two-Level Approach

The main idea of our proposed method is to develop a scalable and modular design by using a *two-level* approach that replaces solving an optimization problem for the entire Chimera graph with solving several similar types of optimization problems, each limited to a single unit cell of the Chimera graph.

Specifically, on the higher level, we design specifications for several types of cell behavior, determined by the properties of the QUBOs defined in these cells, which we call *gadgets*. If the gadgets are arranged in a specific pattern to cover the entire Chimera graph or portions of it, then they produce a QUBO that implements the constraint of interest.

On the lower level, we show that gadgets with the specified properties can actually be designed by solving a corresponding optimization problem. In the next section, we provide details on the implementation of that approach.

## 4   Implementation

### 4.1   Gadgets

As mentioned above, gadgets are QUBO quadratic forms defined on a single Chimera-graph cell. We will arrange these gadgets to cover a rectangular portion $R$ of the Chimera graph. We define three types of gadgets depending on where the corresponding cells will be positioned in $R$:

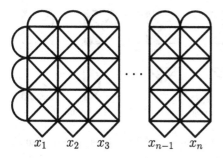  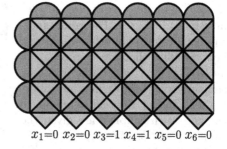

$x_1$  $x_2$  $x_3$    $x_{n-1}$  $x_n$       $x_1=0$ $x_2=0$ $x_3=1$ $x_4=1$ $x_5=0$ $x_6=0$

**Fig. 2.** Arranging the gadgets to solve constraint Eq. (3) in a $k \times n$ region $R$.

**Fig. 3.** Optimal tiling for the inequality $\sum_{i=1}^{6} x_i \leq 3$. (Color figure online)

- *Internal gadget*, denoted by ⊠, for cells (mostly) in the interior of $R$;
- *Problem gadget*, denoted by ▽, for cells on the bottom boundary of $R$, each of which represents a problem variable from $\boldsymbol{x}$; and
- *Boundary gadget*, for cells on the left and top boundaries of region $R$. Depending on the orientation of this gadget, it will be denoted by either ◁ or ◠.

### 4.2  Arranging the Gadgets to Cover the Chimera Graph

As illustrated by Fig. 2, the gadgets cover a rectangular region of the Chimera graph, with problem gadgets on the bottom, boundary gadgets on the top and left boundaries, and internal gadgets on the rest. Moreover, problem variables $\boldsymbol{x}$ are positioned in the problem gadgets, one per gadget. We refer to this QUBO formulation as $Qb(R)$. In the next subsections we will show that $Qb(R)$ is a correct implementation of the constraint specified by Eq. (3) and estimate the gap.

### 4.3  Tiles

In this section we define *tiles*, which are used to easily identify when an assignment to the variables of QUBO problems of a specific kind is an optimal one, merely by examining their types and colors. Specifically, if all the tiles are of "good" type (as defined below) and the adjacent tiles' neighboring sides are colored with the opposite color, then the assignment is an optimal one. Next, we give a more formal description and analysis.

While a gadget refers to the QUBO formulation of a cell, i.e., to the values given to the coefficients $\boldsymbol{a}$ and $\boldsymbol{b}$ in the cell, a *tile* is defined both by the underlining gadget as well as by the values of the cell's variables, i.e., by $\boldsymbol{x}$ and $\boldsymbol{t}$. The combination of the tile types of all the cells of $R$ is called a *tiling* of $R$. In order to estimate the value of a given variable assignment, i.e., a tiling of $R$, we categorize the tiles as *good*, meaning that the value $\alpha$ of the cell's QUBO is the smallest possible for any cell, and *bad*, meaning that the corresponding QUBO value exceeds that minimum by at least a parameter $\gamma > 0$, which we call a *gap*. There are 8 types of good tiles illustrated in Fig. 4. The colors indicate values of the interface variables, i.e., variables that are connected by an active (i.e., with nonzero bias) coupler to a variable in a neighboring cell. (In fact, in order to maximize the gap, such couplers are given the value $-1$, the smallest value allowed by the hardware.) Specifically, red

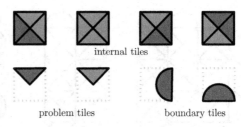

**Fig. 4.** The set of good tiles. (Color figure online)

corresponds to an interface variable value of 1, and green corresponds to 0. Crucially, the colors are used to characterize the interactions between neighboring tiles as described next.

Similarly to the good and bad tile types, we define *good* interactions when the colors on both sides of the common border are different, and *bad*, if they are the same. If the interaction is good, then the value of the quadratic term corresponding to the active coupler connecting the tiles will be negative, generating a reward (i.e., reducing the value of the QUBO, which is beneficial for a minimization problem), and if it is bad, then that value is positive, generating a penalty (increasing the value of the QUBO). We require that the difference between a reward and the smallest penalty for good/bad interactions be $\gamma$, the same as the gap between good and bad tiles.

The tiling of $R$ is called good, if all the tiles and tile interactions are good. The main property of the construction defined in Sect. 4.2 is that there exists a good tiling of $R$ if and only if the values of the problem variables $x$ satisfy the constraint Eq. (3). From this property, it follows that if the constraint is not satisfied, then there is at least one bad tile or connection, and therefore, the gap for $Qb(R)$ between variables satisfying and not satisfying Eq. (3) is $\gamma$.

We will illustrate why this property holds by an example, the formal proof is similar to the one given in our prior report [14].

Figure 3 shows a good tiling for the constraint $\sum_{i=1}^{6} x_i \leq 3$ implemented using gadgets as illustrated in Fig. 2. Note that all boundary tiles have to be red in a good tiling, since this is the only tile type available for a boundary gadget. Furthermore, the colors of the problem tiles encode the variables $x$; in our case, ▽ for $x_1 = x_2 = x_5 = x_6 = 0$ and ▽ for $x_3 = x_4 = 1$. We show that the types of the remaining (internal) tiles are uniquely determined in a good tiling. Notice that each internal tile consists of four triangles, and we are looking at the color of the top one. The main property is that that the number of red colored top triangles decreases by one when we go row by row up from the bottom to the top, until reaching zero. Specifically, the first row from the bottom, consisting of problem tiles, has two red triangles ▽, on position 3 and 4, corresponding to the two $+1$ $x$ values. The second row has only one good way to be tiled: going from left to right, we can only tile it by a sequence of ⊠ tiles, followed by one ⊠ tile above the leftmost problem red ▽ tile, one ⊠ tile above the other problem red ▽ tile, and ending with a sequence of ⊠ tiles. As a result, that row has on its top boundary one fewer red top-triangle than the previous one, since the $x_3$ and $x_4$ tiles ▼ were replaced by one ⊠ (green top) and one ⊠ (red top).

Similarly, the remaining red top-triangle tile ⊠ is eliminated by another ⊠ in the next row. In general, if there are $l$ red problem ▼ tiles in the bottom row, in the next $i$ rows there are exactly $l - i$ red top-triangle tiles (all of ⊠ type) for $1 \leq i \leq l$, assuming $l \leq k$.

The remaining $k - l$ rows can only be tiled by sequences of ⊠ internal tiles with green top-triangles. (In our example $k - l = 1$, but it is clear that one can put any number of rows consisting of ⊠ internal tiles on top of each other.) The red boundary tiles ◭ on the last row guarantee that $l \leq k$ in a good tiling, since the row below it should have all red-top tiles being eliminated.

### 4.4 Solving the Optimization Problems

In the previous section we showed that if tiles with properties as described (i.e. for each gadget, the QUBO values of good and bad tiles for this gadget have a gap $\gamma$) then the QUBO formulation described in Sect. 4.2 implements the constraint Eq. (3) with gap $\gamma$. In order to show that such tiles do exist, for each gadget, we need to solve an optimization problem similar to Problem (i')-(ii'), but restricted to a single cell. Despite being a mixed-integer programming (MIP) problem, it is of small size (16 binary variables and 16 constraints), and easily solvable by modern solvers (in a fraction of a second by Gurobi [9]).

## 5 Analysis

We implemented the tiling scheme described in Sect. 3 using Python and D-Wave Systems's Ocean libraries [6]. We now present an empirical evaluation of this implementation.

As our initial test, we evaluate the $\{k = 3, n = 8\}$ case (i.e., 3 of 8 bits set to 1), comparing measurements taken on Google/NASA/USRA's D-Wave 2000Q quantum annealer with those acquired from a classical simulated-annealing algorithm (implemented in Ocean). We begin with a comparison of correctness. Recall that all annealers are probabilistic and that correct output, corresponding to finding a global minimum, is not guaranteed.

Figure 5 plots a histogram of the results of 1,000,000 anneals. The horizontal axis corresponds to the count of variables that were measured as 1 when the anneal completed, and the vertical axis indicates the number of observations of that count. Ideally, there should be a single bar of height 1,000,000 at position 3 on the $x$ axis. The runs performed using simulated annealing have the correct mode of 3, observed on 35% of the anneals. Disappointingly, the mode of the D-Wave measurements is 5, not 3. 42% of the anneals set five variables to 1, while only 6% set three variables to 1. However, the correctness improved when we ran the D-Wave in virtual full-yield Chimera (VFYC) mode, which includes a classical post-processing step that nudges near-optimal solutions towards optimal

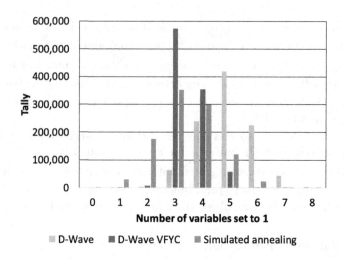

**Fig. 5.** Comparison of quantum annealing to simulated annealing for $\{k = 3, n = 8\}$

solutions. This case performed better than both the raw D-Wave runs and the simulated-annealing runs, observing the correct mode and seeing 58% of the anneals setting three variables to 1. We do not yet know why the raw D-Wave performed so poorly on this test but are currently investigating the issue.

We now turn our attention to execution time. Table 1 lists the time in seconds required to perform 10,000 anneals, averaged over 10 trials. The annealing time for a D-Wave 2000Q is user-specified and ranges from 1μs to 2000 μs. We retained the default annealing time of 20 μs for our experiments. Simulated annealing was run on a workstation

**Table 1.** Execution time for 10,000 anneals, averaged over 10 trials

| Annealer | Time (s) | (±%) |
|---|---|---|
| D-Wave | 17.2 ± 1.7 | (9.9%) |
| D-Wave VFYC | 19.1 ± 2.6 | (13.5%) |
| Simulated annealing | 101.5 ± 1.1 | (1.1%) |

containing two 3.0 GHz Intel Broadwell processors (E5-2687WV4) and 64 GB of DRAM. As Table 1 indicates, the D-Wave is over five times faster than Ocean's implementation of simulated annealing running on a workstation. However, classical overheads—time to transfer the QUBO 1000 miles/1600 km (and back) over an SSL-encrypted network connection, time for the job scheduler to schedule the execution, time for the D-Wave to compute the analog waveforms corresponding to the digital QUBO, etc. [7]—dominate the time spent in quantum annealing. Specifically, at 10,000 anneals of 20 μs apiece, only 0.2 s of the 17.2 s or 19.1 s listed in the table (just over 1%) are spent in annealing proper.

# 6   Conclusions

We developed an optimization approach for embedding penalty constraints of linear inequality type for quantum annealing and applied it to the Chimera graph topology used by the current D-Wave systems. Our approach enables inequality constraints of the type shown in Eq. (3) to be implemented as a QUBO of $O(nk)$ variables. Such QUBOs can be embedded into a Chimera graph without chains (i.e., the QUBO structure is consistent with the Chimera-graph topology). The time for finding such an embedding is independent of the number of variables and depends only on the number of qubits in a cell. The experimental analysis shows that solving the resulting QUBO problem with a classical solver or with D-Wave plus classical postprocessing is accurate, but solutions with a purely quantum solver lack sufficient accuracy. In our future research, we plan to investigate the sources of such inaccuracy and will be looking for ways to increase the robustness.

We note that the construction from Fig. 2 can also be used to solve interval constraints of the type $k_1 \leq \sum_{i=1}^{n} x_i \leq k_2$. To see that, notice that the problem variables are on the bottom boundary of the region. We can therefore create a new construction by merging the current one from Fig. 2 with parameter $k$ set to $k_2$ and a mirror construction with parameter $k$ set to $n - k_1$ and reusing the same problem tiles. Clearly, the top half will implement the inequality $\sum_{i=1}^{n} x_i \leq k_2$, and we want the bottom one to implement the inequality $k_1 \leq \sum_{i=1}^{n} x_i$. To achieve that, we need to make the lower portion "count" the green ($x_i = -1$) tiles rather than the red ones, which can be done by a slight modification of the tiles in the bottom half (treating red color as green and vice versa). As a result, the bottom half will implement the inequality $\sum_{i=1}^{n}(1 - x_i) \leq n - k_2$ , which is equivalent to $n - \sum_{i=1}^{n} x_i \leq n - k_2$ or $\sum_{i=1}^{n} x_i \geq k_2$.

**Acknowledgments.** Research presented in this article was supported by the Laboratory Directed Research and Development program of Los Alamos National Laboratory under project numbers 20180267ER and 20190065DR. This work was also supported by the U.S. Department of Energy through Los Alamos National Laboratory. Los Alamos National Laboratory is operated by Triad National Security, LLC for the National Nuclear Security Administration of the U.S. Department of Energy (contract no. 89233218CNA000001).

# References

1. Barahona, F.: On the computational complexity of Ising spin glass models. J. Phys. A: Math. Gen. **15**(10), 3241 (1982)
2. Bian, Z., Chudak, F., Israel, R., Lackey, B., Macready, W.G., Roy, A.: Discrete optimization using quantum annealing on sparse Ising models. Front. Phys. **2**, 56:1–56:10 (2014)
3. Boothby, T., King, A.D., Roy, A.: Fast clique minor generation in Chimera qubit connectivity graphs. Quantum Inf. Process. **15**(1), 495–508 (2016)
4. Bunyk, P.I., et al.: Architectural considerations in the design of a superconducting quantum annealing processor. IEEE Trans. Appl. Supercond. **24**(4), 1–10 (2014)

5. Choi, V.: Minor-embedding in adiabatic quantum computation: I. The parameter setting problem. Quantum Inf. Process. **7**(5), 193–209 (2008)
6. D-Wave Systems Inc.: D-Wave's Ocean Software. https://ocean.dwavesys.com/
7. D-Wave Systems Inc.: Measuring computation time on D-Wave systems. User Manual 09–1107A-G, 2 February 2018
8. Gibney, E.: D-Wave upgrade: how scientists are using the world's most controversial quantum computer. Nature **541**(7638), 447–448 (2017)
9. Gurobi Optimization Inc.: Gurobi optimizer reference manual (2015)
10. Kadowaki, T., Nishimori, H.: Quantum annealing in the transverse Ising model. Phys. Rev. E **58**, 5355–5363 (1998)
11. Kirkpatrick, S., Gelatt, C.D., Vecchi, M.P.: Optimization by simulated annealing. Science **220**(4598), 671–680 (1983)
12. Kochenberger, G., et al.: The unconstrained binary quadratic programming problem: a survey. J. Comb. Optim. **28**(1), 58–81 (2014)
13. Lucas, A.: Ising formulations of many NP problems. Front. Phys. **2**, 5:1–5:15 (2014)
14. Vyskocil, T., Djidjev, H.: Optimization approach to constraint embedding for quantum annealers. Technical report LA-UR-18-30971, Los Alamos National Laboratory (2018)
15. Vyskocil, T., Djidjev, H.: Simple constraint embedding for quantum annealers. In: International Conference on Rebooting Computing (2018, to appear)

# Assessing Solution Quality of 3SAT on a Quantum Annealing Platform

Thomas Gabor[1]([⊠]), Sebastian Zielinski[1], Sebastian Feld[1], Christoph Roch[1], Christian Seidel[2], Florian Neukart[3], Isabella Galter[2], Wolfgang Mauerer[4], and Claudia Linnhoff-Popien[1]

[1] LMU Munich, Munich, Germany
thomas.gabor@ifi.lmu.de
[2] Volkswagen Data:Lab, Munich, Germany
[3] Volkswagen Group of America, San Francisco, USA
[4] OTH Regensburg/Siemens Corporate Research, Regensburg, Germany

**Abstract.** When solving propositional logic satisfiability (specifically 3SAT) using quantum annealing, we analyze the effect the difficulty of different instances of the problem has on the quality of the answer returned by the quantum annealer. A high-quality response from the annealer in this case is defined by a high percentage of correct solutions among the returned answers. We show that the phase transition regarding the computational complexity of the problem, which is well-known to occur for 3SAT on classical machines (where it causes a detrimental increase in runtime), persists in some form (but possibly to a lesser extent) for quantum annealing.

**Keywords:** Quantum computing · Quantum annealing · D-wave · 3SAT · Boolean satisfiability · NP · Phase transition

## 1 Introduction

Quantum computers are an emerging technology and still subject to frequent new developments. Eventually, the utilization of intricate physical phenomena like superposition and entanglement is conjectured to provide an advantage in computational power over purely classical computers. As of now, however, the first practical breakthrough application for quantum computers is still sought for. But new results on the behavior of quantum programs in comparison to their classical counterparts are reported on a daily basis.

Research in that area has cast an eye on the complexity class NP: It contains problems that are traditionally (and at the current state of knowledge regarding the P vs. NP problem) conjectured to produce instances too hard for classical computers to solve exactly and deterministically within practical time constraints. Still, problem instances of NP are also easy enough that they can be executed efficiently on a (hypothetical) non-deterministic computer.

© Springer Nature Switzerland AG 2019
S. Feld and C. Linnhoff-Popien (Eds.): QTOP 2019, LNCS 11413, pp. 23–35, 2019.
https://doi.org/10.1007/978-3-030-14082-3_3

The notion of computational complexity is based on classical computation in the sense of using classical mechanics to describe and perform automated computations. In particular, it is known that in this model of computation, simulating quantum mechanical systems is hard. However, nature itself routinely "executes" quantum mechanics, leading to speculations [20] that quantum mechanics may be used to leverage greater computational power than systems adhering to the rules of classical physics can provide.

Quantum computing describes technology exploiting the behavior of quantum mechanics to build computers that are (hopefully) more powerful than current classical machines. Instead of classical bits $b \in \{0, 1\}$ they use qubits $q = \alpha |0\rangle + \beta |1\rangle$ where $\alpha$, $\beta$, $|\alpha|^2 + |\beta|^2 = 1$, are probability amplitudes for the basis states $|0\rangle, |1\rangle$. Essentially, a qubit can be in both states 0 and 1 at once. This phenomenon is called superposition, but it collapses when the actual value of the qubit is measured, returning either 0 or 1 with a specific probability and fixing that randomly acquired result as the future state of the qubit. Entanglement describes the effect that multiple qubits can be in superpositions that are affected by each other, meaning that the measurement of one qubit can change the assigned probability amplitudes of another qubit in superposition. The combination of these phenomena allows qubits to concisely represent complex data and lend themselves to efficient computation operations.

The technological platform of quantum annealing is (unlike the generalized concept of quantum computing) not capable of executing general quantum-mechanical computations, but is within current technological feasibility and available to researchers outside the field of quantum hardware. The mechanism specializes in solving optimization problems and can (as a trade-off) work larger amounts of qubits in a useful way than current quantum-mechanically complete platforms.

In this paper, we evaluate the performance of quantum annealing (or more specifically, a D-Wave 2000Q machine) on the canonical problem of the class NP, propositional logic satisfiability for 3-literal clauses (3SAT) [14]. As we note that there is still a remarkable gap between 3SAT instances that can be put on a current D-Wave chip and 3SAT instances that even remotely pose a challenge to classical solvers, there is little sense in comparing the quantum annealing method to classical algorithms in this case (and at this early point in time for the development of quantum hardware). Instead, we are interested in the scaling behavior with respect to problem difficulty. Or more precisely: We analyze if and to what extent quantum annealing's performance suffers under hard problem instances (like classical algorithms do).

We present a quick run-down of 3SAT and the phenomenon of phase transitions in Sect. 2 and continue to discuss further related work in Sect. 3. In Sect. 4 we describe our experimental setup and then present the corresponding results in Sect. 5. We conclude with Sect. 6.

## 2   Preliminaries

Propositional logic satisfiability (SAT) is the problem of telling if a given formula in propositional logic is satisfiable, i.e., if there is a assignment to all involved Boolean variables that causes the whole formula to reduce to the logical value *True*. As such, the problem occurs at every application involved complex constraints or reasoning, like (software) product lines, the tracing of software dependencies or formal methods.

It can be trivially shown that (when introducing a linear amount of new variables) all SAT problems can be reduced to a specific type of SAT problem called 3SAT, where the input propositional logic formula has to be in conjunctive normal form with all of the disjunctions containing exactly three literals.

For example, the formula $\Psi = (x_1 \vee x_2 \vee x_3) \wedge (\neg x_1 \vee x_2 \vee x_3)$ is in 3SAT form and is satisfiable because the assignment $(x_1 \mapsto True, x_2 \mapsto True, x_2 \mapsto True)$ causes the formula to reduce to *True*. The formula $\Phi = (x_1 \vee x_1 \vee x_1) \wedge (\neg x_1 \vee \neg x_1 \vee \neg x_1)$ is also in 3SAT form but is not satisfiable.

**Definition (3SAT).** A 3SAT instance with $m$ clauses and $n$ variables is given as a list of clauses $(c_k)_{0 \leq k \leq m-1}$ of the form $c_k = (l_{3k} \vee l_{3k+1} \vee l_{3k+2})$ and a list of variables $(v_j)_{0 \leq j \leq n-1}$ so that $l_i$ is a literal of the form $l_i \in \bigcup_{0 \leq j \leq n-1} \{v_j, \neg v_j\}$. A given 3SAT instance is *satisfiable* iff there exists a variable assignment $(v_j \mapsto b_j)_{0 \leq j \leq n-1}$ with $b_j \in \{True, False\}$ so that $\bigwedge_{0 \leq k \leq m-1} c_k$ reduces to *True* when interpreting all logical operators as is common. The problem of deciding whether a given 3SAT instance is satisfiable is called 3SAT.

3SAT is of special importance to complexity theory as it was the first problem which was shown to be NP-complete [14]. This means that every problem in NP can be reduced to 3SAT in polynomial time. It follows that any means to solve 3SAT efficiently would thus give rise to efficient solutions for any problem in NP like graph coloring, travelling salesman or bin packing.

Despite the fact that for NP-complete problems in general no algorithm is known that can solve all problem instances of a problem efficiently (i.e., in polynomial time), it is within the scope of knowledge that "average" problem instances of many NP-complete problems, including 3SAT, are easy to solve [10]. In Ref. [37] this characteristic is described with a phase transition. The boundary of the phase transition divides the problem space into two regions. In one region, a solution can be found relatively easily, because the solution density for these problems is high, whereas in the other region, it is very unlikely that problems can contain a correct solution at all. Problems that are very difficult to solve are located directly at this phase boundary [10].

It can be observed that, with randomly generated 3SAT instances, the probability of finding a correct solution decreases abruptly when the ratio of clauses to variables $\alpha = m/n$ exceeds a critical value of $\alpha_c$ [36]. According to [35] this critical point is $\alpha_c \approx 4.267$ for randomly generated 3SAT instances. In the surrounding area of the critical point, finding a solution (i.e., deciding if the instance is satisfiable) is algorithmically complex. Figure 1 illustrates this phenomenon.

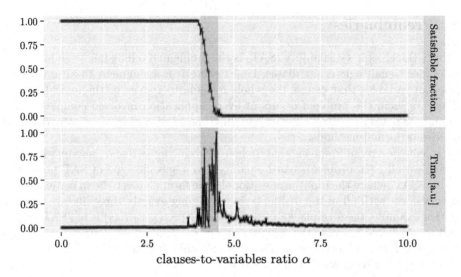

**Fig. 1.** Phase transition of 3SAT. The bottom plot shows the computational time required to determine satisfiability of randomly chosen 3SAT instances with a specific clauses-to-variables ratio $\alpha$ on a standard solver. The area around the critical point $\alpha_c \approx 4.267$ is shaded in blue. The upper portion shows the probability that instances with a particular ratio $\alpha$ are solvable. In the region around the critical point, it is hard to determine whether a problem instance can be fulfilled with a concrete allocation or not. (Color figure online)

To assess the solution quality of randomly generated 3SAT instances we generate instances in every complexity region. The results are discussed in Sect. 5.

## 3   Related Work

It is one of the cornerstones of complexity theory that solving NP-complete decision problems is strongly believed to be not efficiently possible [14,40]. Any NP-complete problem can also be cast as an optimization problem, which allows for employing well-known optimization algorithms to find approximate solutions—typical methods include tabu search [22,23] and simulated annealing [11,27]. Countless other efficient approximation methods, together with an elaborate taxonomy on approximation quality (how much does a given solution differ from a known global optimum?) and computational effort (how many time steps are required until an approximate solution that satisfies given quality goals is available?), have been devised [6].

An intriguing connection that has received substantial attraction exists between (computational) NP-complete problems and the (physical) concept of phase transitions, as detailed in Sect. 2. First investigations of the phenomenon have been performed by Kirkpatrick et al. [28]; Monasson et al. first suggested a connection between the type of phase transition and the associated computational costs of a problem [37]. From the abundant amount of more recent investigations, we would like to highlight the proof by Ding et al. [16] that establishes a threshold value for the phase transition. Our work benefits from the above insights by selecting the "most interesting", i.e., computationally hardest, scenarios as investigation target.

The idea of obtaining solutions for NPO (NP optimization) problems by finding the energy ground state (or states) of a quantum mechanical system was used, for instance, by Apolloni et al. [4,5] to solve combinatorial optimization problems. The general idea of quantum annealing has been independently rediscovered multiple times [2,3,21,26].

Quantum annealing techniques are usually applied to solving NP-complete decision problems, or optimization problems from class NPO. Lucas [30] reviews how to formulate a set of key NP problems in the language of adiabatic quantum computing, i.e., quadratic unconstrained binary optimization (QUBO). In particular, problems of the types "travelling salesman" or "binary satisfiability" that are expected to have a major impact on practical computational applications if they can be solved advantageously on quantum annealers have undergone a considerable amount of research [7,24,39,41,44,46]. Further effort has been made on combining classical and quantum methods on these problems [19].

Comparing the computational capabilities of classical and quantum computers is an intriguing and complex task, since the deployed resources are typically very dissimilar. For instance, the amount of instructions required to execute a particular algorithm is one of the main measures of efficiency or practicability on a classical machine, whereas the notion of a discrete computational "step" is hard to define on a quantum annealing device. Still, multiple efforts have been made towards assessing quantum speedup [42,43]. For the quantum gate model, a class of problems exhibiting quantum speedup has been found lately [9]. Interest in quantum computing has also spawned definitions of new complexity classes (e.g., [29,38]), whose relations to traditional complexity classes have been and are still subject to ongoing research [8,32].

These questions hold regardless of any specific physical or conceptual implementation of quantum computing since their overall computational capabilities are known to be largely interchangeable; for instance, McGeoch [33] discusses the equivalence of gate-based and adiabatic quantum computing. Consequently, our work focuses not on comparing quantum and classical aspects of solving particular problems, but concentrates on understanding peculiarities of solving one particular problem (3SAT, in our case) in-depth.

Formulating 3SAT problems on a quantum annealing hardware has been previously considered [12,13,18], and we rely on the encoding techniques presented there. Van Dam [45] and Farhi [17] have worked on analyzing the complexity of solving general 3SAT problems. Hsu et al. have considered the complexity-wise easier variation 2SAT as a benchmarking problem to compare various parameter configurations of their quantum annealer [25].

# 4    Experimental Setup

Quantum annealing is an optimization process that can be implemented in hardware. It is built upon the adiabatic theorem that provides conditions under which an initial ground-state configuration of a system evolves to the ground state of another configuration that minimizes a specific user-defined energy function [33]. As in the real world the required conditions for the theorem can only be approximated, the results of quantum annealing are not deterministically optimal but show a probabilistic distribution, ideally covering the desired optimal value.

D-Wave's quantum annealer is the first commercial machine to implement quantum annealing. Its interface is built on two equivalent mathematical models for optimization problems called Ising and QUBO, the latter of which will be used for the work of this paper. Quadratic Unconstrained Binary Optimization (QUBO) problems can be formulated as a quadratic matrix $Q_{ij}$. Quantum annealing then searches for a vector $x \in \{0,1\}^n$ so that $\sum_i \sum_{j<i} Q_{ij} x_i x_j + \sum_i Q_i x_i$ is minimal. The promise of quantum annealing is that—using quantum effects—specialized hardware architectures are able to solve these optimization problems much faster than classical computers in the future.

The main goal of this paper is to analyze the inherently probabilistic distribution of return values generated by quantum annealing when trying to solve hard optimization problems. We choose to demonstrate such an analysis on 3SAT because it is the canonical problem of the class NP, which is a prime target for research on performance improvements via quantum technology with respect to classical computers [30,34].

## 4.1    Defining 3SAT as a QUBO

3SAT is usually not formulated as an optimization problem (see Sect. 2), or defined by an equivalent QUBO problem, as is required by the annealer. Thus, we require a (polynomial-time) translation of any 3SAT instance into a QUBO so that the solutions generated by the quantum annealer can be translated back to solutions of the initial 3SAT instance.

Following [12,13], we translate 3SAT into the Weighted Maximum Independent Set (WMIS) problem and then translate the WMIS instance into a QUBO (we find it convenient to specify the polynomial coefficients in matrix form). We omit the details of this process and instead refer to *op. cit.* and Lucas [30]. However, we shall briefly discuss the implications of the translation process.

A 3SAT instance, that is, a formula with $m$ clauses for $n$ variables, requires a QUBO matrix of size $3m \times 3m$ with the solution vector $x \in \{0,1\}^{3m}$. The solution can be thought of as using a qubit for each literal in the initial formula and thus consisting of a triplet of qubits for each 3SAT clause. This usually means that we have much more qubits than variables in the formula. Nonetheless, a QUBO solution is mapped to a value assignment for the variables in the 3SAT formula. Thus, when running successfully, the quantum annealer will output a satisfying assignment for a given 3SAT formula. We can check if the assignment really is correct (i.e., each variable has a value assigned and the whole formula reduces to *True*) using few instructions of classical computation. Obviously, if among several experimental runs the quantum annealer does return just one correct assignment, the corresponding 3SAT formula is satisfiable. If the quantum annealer only returns incorrect assignments, we will regard the formula as unsatisfiable (although the prove of that is only probabilistic).

There are some aspects to note about how the QUBO solution vectors are mapped to variable assignments. Given a QUBO solution vector $(x_i)_{0 \leq i \leq 3m-1}$ for a 3SAT formula with literals $(l_i)_{0 \leq i \leq 3m-1}$, a variable $v$ is assigned the value *True* if it occurs in a literal $l_i = v$ and $x_i = 1$. Likewise, a variable $v$ is assigned the value *False* if it occurs in a literal $l_i = \neg v$ and $x_i = 1$. It is important to note that $x_i = 0$ has *no implication* on the value of the variable in $l_i$.

Intuitively, we can interpret $x_i = 1$ to mean "use the value of $l_i$ to prove the satisfaction of clause $c_{(i \mod 3)}$". From our QUBO optimization, we expect to find one (and only one) suitable $l_i$ for every clause in the 3SAT formula.[1]

This is important as it opens up a wide range of different QUBO solutions which may just encode the exact same variable assignment at the 3SAT level. However, it also means that seemingly suboptimal QUBO solutions may encode correct 3SAT assignments. For example, consider the (a little redundant) 3SAT formula $(v_0 \vee v_1 \vee v_2) \wedge (v_0 \vee v_1 \vee v_2)$: The QUBO solution $x = 100001$ would imply the assignment of $v_0 = True$ and $v_2 = True$, which indeed is theoretically sufficient to prove the formula satisfiable. The exact same assignment would be implied by $x = 001100$. However, note that none of these imply a full assignment of every variable in the 3SAT instance since none say anything about the value of $v_1$. Still, we can trivially set $v_1$ to any arbitrary value and end up with a correct assignment. Also note that while the QUBO is built in such a way to opt for one single value 1 per triplet in the bit string, even bitstrings violating this property can encode correct solution. In our example, the suboptimal QUBO solution $x = 100000$ still encodes all necessary information to prove satisfiability.

## 4.2   Evaluating Postprocessing

As can be seen from the last example, postprocessing is an integral part of solving problems with quantum annealing. As discussed earlier in this section, we consider a QUBO solution correct, if it not only matches the expected structure for minimizing the QUBO energy function, but instead iff it directly implies a correct assignment in the definition given above. Thus, while the expected structure for QUBO optimizes $x$ so that the amount bits $x_i$ assigned 1 equals the amount of clauses $m$, we also consider less full answers correct.

On top of that, there are solutions that cannot be mapped to an assignment immediately, but still with almost no effort. We want to regard these as well and implemented a postprocessing step we call *logical postprocessing*. It is applied whenever none of the qubits corresponding to a single clause $c_k$ are set to 1 by

---

[1] This intuition matches the concept of constructivism in logic and mathematics. We are not only looking for the correct answer, but are looking for a correct and complete proof of an answer, giving us a single witness for each part of the formula.

the quantum annealer and the respective QUBO solution is not already correct. In that case, we iterate through all literals $l_i$ in that clause $c_k$ and check if we could set $x_i = 1$ without contradicting any other assignment made within $x$. If we find such an $l_i$, we set $x_i = 1$ and return the altered bitstring $x$.

The software platform provided by D-Wave to use the quantum annealer already offers integrated postprocessing methods as well, which we will also empirically show to be more powerful than logical postprocessing in the following Sect. 5. Again, for greater detail we refer to the D-Wave documentation on that matter [15]. At a glance, the employed postprocessing method splits the QUBO matrix into several subproblems, tries to optimize these locally, and then integrates that local solution into the complete solution if it yields an improvement. We call this method *D-Wave postprocessing*.

To evaluate the solution quality regarding 3SAT, we employ both methods. The goal is to assess expected quality on a 3SAT-to-3SAT level, i.e., we measure how well we can solve the given 3SAT instance and regard the translation to and from QUBO as a mere technical problem that is not of interest for this paper.

## 5  Evaluation

To assess the solution quality of 3SAT on a quantum annealing platform, using the previously discussed method of encoding 3SAT problems, we ran several experiments on a D-Wave 2000Q system. Using ToughSAT[2] we generated 3SAT instances of various difficulty (i.e., with various values for $\alpha$). However, as discussed in Sect. 2, for $|\alpha - 4.2| \gg 0$ problem instances become very easy to solve. We observed that effect on the quantum annealer as well, since *all* of these instances were easily (i.e. 100% of the time) solved on the D-Wave machine. Thus, for the remainder of this section, we focus on hard instances (approximated by $\alpha = 4.2$) to assess solution quality in the interesting problem domain.

Experiments have shown that using the standard embedding tools delivered with the D-Wave platform, we can only reliably find a working embedding on the D-Wave 2000Q chip for 3SAT instances with at most 42 clauses [1]. To maintain $\alpha \approx 4.2$, the generated 3SAT instances contain 10 different variables. We only assess solution quality for 3SAT instances that are satisfiable, but do not provide this information to the solver.

Figure 2 shows the result distribution of these runs on the D-Wave machine. On the x-axis, we sorted the returned results according to the bits that have been assigned the value 1 or *True*. As discussed in Sect. 4 the optimal solution is supposed to set one bit for each clause, i.e., is supposed to contain 42 bits set to *True*. However, as there are only 10 different variables, there theoretically exist answers that only set 10 bits but that still map to a complete and valid solution for the given 3SAT instance. From Fig. 2 we can see that some of these solution are found for bitcounts starting from 37 through 41. Interestingly, the complete range of answers gathered seems to follow a distribution centered around 37 or

---

[2] https://toughsat.appspot.com/.

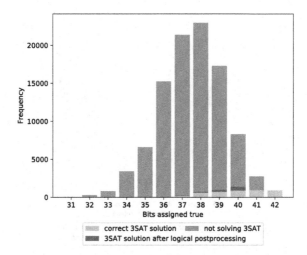

**Fig. 2.** Distribution of correct (green) and incorrect (red) answers returned by the quantum annealer *without D-WAVE postprocessing*. Answers that can trivially be transformed into valid answers using logical postprocessing are marked in yellow. The plot shows 100,000 answers in total for 100 different hard 3SAT instances ($\alpha \approx 4.2$). (Color figure online)

38 and no answers with more than 42 bits are returned. This means that the constraint of never setting multiple bits per clause is fully respected in the evaluation of our QUBO matrix. Note that although there are 5,283 correct solutions in total, these are only distributed across 24 of the 100 randomly generated problem instances. Thus, most of them have not been solved at all.

Furthermore, we applied the logical postprocessing described in Sect. 4 to the incorrect answers in Fig. 2. However, it shows little improvement on the total amount of correct answers collected. We expect the postprocessing method delivered with the D-Wave software package to be more powerful as it runs local search along more axes of the solution space than the logical postprocessing does. So we ran the complete evaluation experiment again, only this time turning on the integrated postprocessing. The results are shown in Fig. 3.

We observed that the D-Wave postprocessing managed to optimize all correct but "incomplete" answers, mapping them to a solution with 42 bits assigned the value *True*. Out of the 100,000 queries, this yielded 25,142 correct answers. Moreover, these correct answers span 99 of the 100 randomly generated 3SAT instances so that we consider the problem solved. Effectively, this shows that quantum annealing does suffer from a breakdown in expected solution quality at the point of the phase transition in the 3SAT problem. In comparison to the immense decrease in performance seen in classical solvers (cf. Sect. 2), a drop to around 25% precision (if it was to persist on larger chip sizes) appears rather desirable, though. A quick example: To achieve a $1 - 10^{-12}$ confidence of returning the correct answer our experimental setup requires around 97 queries.

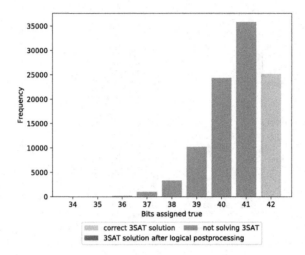

**Fig. 3.** Distribution of correct (green) and incorrect (red) answers returned by the quantum annealer *using D-WAVE postprocessing*. Answers that can trivially be transformed into valid answers using logical postprocessing are marked in yellow. The plot shows 100,000 answers in total for 100 different hard 3SAT instances ($\alpha \approx 4.2$). (Color figure online)

At a glance, that scaling factor with respect to problem difficulty is much better than what is observed for classical algorithms: For example, in the data used for Fig. 1 we observed performance decrease up to one order of magnitude larger. It is important to note, however, that these experiments were performed for problem instances so small that their evaluation does not pose a challenge to classical processors at all, i.e., below the point of reasonable performance metrics. Thus, these results only proof relevant to practical applications if they scale with future versions of quantum annealing hardware that can tackle much larger problem instances.

So far, we have not discerned between different correct solutions. We were content as long as the algorithm returned but one. However, for the user it is interesting to know if he or she will receive the same solution with every answer or an even distribution across the complete solution space. Our experiments show that when a lot of correct solutions are found for a certain problem instance, there are cases where we can see a clear bias towards a specific solution variant. Figure 4 shows the distributions of specific solutions for formulae that yielded many solutions even when evaluated without any postprocessing. While some formulae seem to yield rather narrow distributions over the different possible answers, others definitely seem to have a bias towards certain solutions. However, the former also tend to have relatively smaller sample sizes as there are less solutions in total to consider. Further investigation could still reveal a distinctive distribution in these cases as well. Thus, we consider this behavior of the quantum annealer to be roughly in line with the findings of [31], who show an exponential bias in ground-state sampling of a quantum annealer.

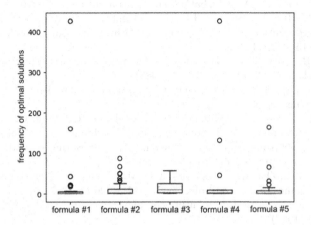

**Fig. 4.** Frequency of occurrence of different solutions for 5 formulae with many returned solutions without any postprocessing. While most solutions are found once or just a few times, there are specific solutions that are found much more often.

## 6   Conclusion

We have shown that problem difficulty of 3SAT instances also affects the performance of quantum annealing as it does for classical algorithms. However, bound by the nature of both approaches, the effects are quite different with complete classical algorithms showing longer runtimes and quantum annealing showing less precision. A first quantification of that loss of precision suggests that it may not be too detrimental and comparatively easy to deal with. However, because of the maximum available chip size for quantum annealing hardware at the moment, no large-scale test could be performed. No real assumptions on the scaling of this phenomenon (and thus the eventual real-world benefit) can be made yet.

Our results suggest there are cases where single solutions from a set of equally optimal solutions are much more likely to be returned than others. This observation is in line with other literature on the results of quantum annealing. However, it is interesting to note that it translates into the original problem space of 3SAT.

The observed results will gain more practical relevance with larger chip sizes for quantum annealers. We thus suggest to perform these and/or similar tests for future editions of quantum annealing hardware. If the effects persist, they can indicate a substantial advantage of quantum hardware over other known approaches for solving NP-complete problems.

**Acknowledgement.** Research was funded by Volkswagen Group, department Group IT.

# References

1. Adams, D.: The Hitchhiker's Guide to the Galaxy (1979)
2. Albash, T., Lidar, D.A.: Adiabatic quantum computing. arXiv:1611.04471 (2016)
3. Amara, P., Hsu, D., Straub, J.E.: Global energy minimum searches using an approximate solution of the imaginary time Schrödinger equation. J. Phys. Chem. **97**(25), 6715–6721 (1993)
4. Apolloni, B., Carvalho, C., De Falco, D.: Quantum stochastic optimization. Stoch. Process. Their Appl. **33**(2), 233–244 (1989)
5. Apolloni, B., De Falco, D., Cesa-Bianchi, N.: A numerical implementation of "quantum annealing". Technical report (1988)
6. Ausiello, G., Protasi, M., Marchetti-Spaccamela, A., Gambosi, G., Crescenzi, P., Kann, V.: Complexity and Approximation: Combinatorial Optimization Problems and Their Approximability Properties, 1st edn. Springer, Heidelberg (1999). https://doi.org/10.1007/978-3-642-58412-1
7. Benjamin, S.C., Zhao, L., Fitzsimons, J.F.: Measurement-driven quantum computing: Performance of a 3-SAT solver. arXiv:1711.02687 (2017)
8. Bernstein, E., Vazirani, U.: Quantum complexity theory. SIAM J. Comput. **26**(5), 1411–1473 (1997)
9. Bravyi, S., Gosset, D., Koenig, R.: Quantum advantage with shallow circuits. Science **362**(6412), 308–311 (2018)
10. Cheeseman, P.C., Kanefsky, B., Taylor, W.M.: Where the really hard problems are. In: IJCAI, vol. 91 (1991)
11. Chen, L., Aihara, K.: Chaotic simulated annealing by a neural network model with transient chaos. Neural Netw. **8**(6), 915–930 (1995)
12. Choi, V.: Adiabatic quantum algorithms for the NP-complete maximum-weight independent set, exact cover and 3SAT problems. arXiv:1004.2226 (2010)
13. Choi, V.: Different adiabatic quantum optimization algorithms for the NP-complete exact cover and 3SAT problems. Quant. Inform. Comput. **11**(7–8), 638–648 (2011)
14. Cook, S.A.: The complexity of theorem-proving procedures. In: Proceedings of the Third Annual ACM Symposium on Theory of Computing. ACM (1971)
15. D-Wave Systems: Postprocessing Methods on D-Wave Systems (2016)
16. Ding, J., Sly, A., Sun, N.: Proof of the satisfiability conjecture for large k. In: Proceedings of the 47th Annual ACM Symposium on Theory of Computing, STOC 2015. ACM, New York (2015)
17. Farhi, E., Goldstone, J., Gosset, D., Gutmann, S., Meyer, H.B., Shor, P.: Quantum adiabatic algorithms, small gaps, and different paths. arXiv:0909.4766 (2009)
18. Farhi, E., Goldstone, J., Gutmann, S., Sipser, M.: Quantum computation by adiabatic evolution. arXiv preprint quant-ph/0001106 (2000)
19. Feld, S., et al.: A hybrid solution method for the capacitated vehicle routing problem using a quantum annealer. arXiv preprint arXiv:1811.07403 (2018)
20. Feynman, R.P.: Simulating physics with computers. Int. J. Theor. Phys. **21**, 467–488 (1982)
21. Finnila, A., Gomez, M., Sebenik, C., Stenson, C., Doll, J.: Quantum annealing: a new method for minimizing multidimensional functions. Chem. Phys. Lett. **219**(5–6), 343–348 (1994)
22. Gendreau, M., Hertz, A., Laporte, G.: A Tabu search heuristic for the vehicle routing problem. Manag. Sci. **40**(10), 1276–1290 (1994)
23. Glover, F., Laguna, M.: Tabu search*. In: Pardalos, P.M., Du, D.-Z., Graham, R.L. (eds.) Handbook of Combinatorial Optimization, pp. 3261–3362. Springer, New York (2013). https://doi.org/10.1007/978-1-4419-7997-1_17

24. Heim, B., Brown, E.W., Wecker, D., Troyer, M.: Designing adiabatic quantum optimization: a case study for the TSP. arXiv:1702.06248 (2017)
25. Hsu, T.J., Jin, F., Seidel, C., Neukart, F., De Raedt, H., Michielsen, K.: Quantum annealing with anneal path control: application to 2-SAT problems with known energy landscapes. arXiv:1810.00194 (2018)
26. Kadowaki, T., Nishimori, H.: Quantum annealing in the transverse Ising model. Phys. Rev. E **58**(5), 5355 (1998)
27. Kirkpatrick, S., Gelatt, C.D., Vecchi, M.P.: Optimization by simulated annealing. Science **220**(4598), 671–680 (1983)
28. Kirkpatrick, S., Selman, B.: Critical behavior in the satisfiability of random Boolean expressions. Science **264**(5163), 1297–1301 (1994)
29. Klauck, H.: The complexity of quantum disjointness. In: Leibniz International Proceedings in Informatics, vol. 83. Schloss Dagstuhl-Leibniz-Zentrum fuer Informatik (2017)
30. Lucas, A.: Ising formulations of many NP problems. Front. Phys. **2**, 5 (2014)
31. Mandra, S., Zhu, Z., Katzgraber, H.G.: Exponentially biased ground-state sampling of quantum annealing machines with transverse-field driving hamiltonians. Phys. Rev. Lett. **118**(7), 070502 (2017)
32. Marriott, C., Watrous, J.: Quantum Arthur-Merlin games. Comput. Complex. **14**(2), 122–152 (2005)
33. McGeoch, C.C.: Adiabatic quantum computation and quantum annealing: theory and practice. Synth. Lect. Quantum Comput. **5**(2), 1–93 (2014)
34. McGeoch, C.C., Wang, C.: Experimental evaluation of an adiabiatic quantum system for combinatorial optimization. In: Proceedings of the ACM International Conference on Computing Frontiers. ACM (2013)
35. Mézard, M., Zecchina, R.: Random k-satisfiability problem: from an analytic solution to an efficient algorithm. Phys. Rev. E **66**(5), 056126 (2002)
36. Monasson, R., Zecchina, R.: Entropy of the k-satisfiability problem. Phys. Rev. Lett. **76**(21), 3881 (1996)
37. Monasson, R., Zecchina, R., Kirkpatrick, S., Selman, B., Troyansky, L.: Determining computational complexity from characteristic "phase transitions". Nature **400**(6740), 133 (1999)
38. Morimae, T., Nishimura, H.: Merlinization of complexity classes above BQP. arXiv:1704.01514 (2017)
39. Moylett, D.J., Linden, N., Montanaro, A.: Quantum speedup of the traveling-salesman problem for bounded-degree graphs. Phys. Rev. A **95**(3), 032323 (2017)
40. Murty, K.G., Kabadi, S.N.: Some NP-complete problems in quadratic and nonlinear programming. Math. Program. **39**(2), 117–129 (1987)
41. Neukart, F., Compostella, G., Seidel, C., von Dollen, D., Yarkoni, S., Parney, B.: Traffic flow optimization using a quantum annealer. Front. ICT **4**, 29 (2017)
42. Rønnow, T.F., et al.: Defining and detecting quantum speedup. Science **345**(6195), 420–424 (2014)
43. Somma, R.D., Nagaj, D., Kieferová, M.: Quantum speedup by quantum annealing. Phys. Rev. Lett. **109**(5), 050501 (2012)
44. Strand, J., Przybysz, A., Ferguson, D., Zick, K.: ZZZ coupler for native embedding of MAX-3SAT problem instances in quantum annealing hardware. In: APS Meeting Abstracts (2017)
45. Van Dam, W., Mosca, M., Vazirani, U.: How powerful is adiabatic quantum computation? In: 42nd IEEE Symposium on Foundations of Computer Science. IEEE (2001)
46. Warren, R.H.: Small traveling salesman problems. J. Adv. Appl. Math. **2**(2), 101–107 (2017)

# Principles and Guidelines for Quantum Performance Analysis

Catherine C. McGeoch[✉]

D-Wave Systems, Burnaby, BC, Canada
cmcgeoch@dwavesys.com

**Abstract.** Expanding access to practical quantum computers prompts a widespread need to evaluate their performance. Principles and guidelines for carrying out sound empirical work on quantum computing systems are proposed. The guidelines draw heavily on classical experience in experimental algorithmics and computer systems performance analysis, with some adjustments to address issues in quantum computing. The focus is on issues related to quantum annealing processors, although much of the discussion applies to more general scenarios.

**Keywords:** Quantum computing · Experimental methodology

## 1 Introduction

There are many reasons for doing experimental work on quantum solvers,[1] including: presenting a proof-of-concept demonstration that a quantum solution is viable; developing empirical models that show how input properties drive performance; identifying "hard" inputs that expose solver vulnerabilities; finding strategies for tuning quantum solvers; and demonstrating a "grand challenge" achievement such as quantum supremacy [8] or quantum speedup [14,26]. Another important reason is *application benchmarking*, defined here as characterizing solver performance in practical use scenarios.

Well-established guidelines for developing and reporting on computational experiments are available from the fields of experimental algorithmics [5–7,9,15,21,22], and computer systems performance analysis [3,4,12,13]. However, merging these guidelines for the quantum scenario is not straightforward, and new rules are needed to address issues specific to quantum computation. This paper presents four principles for experimental study of quantum computers, and several guidelines aligned with these principles, drawing heavily on lessons from classical computing. The focus is on D-Wave quantum annealing processors and optimization performance, although most of the discussion applies generally. Performance is interpreted here to include combinations of runtime, solution quality, or solution sample (i.e. distribution) quality.

---

[1] A solver is an algorithm that has been implemented in code or hardware. We assume throughout that the quantum algorithm and platform are tested as a unit.

© Springer Nature Switzerland AG 2019
S. Feld and C. Linnhoff-Popien (Eds.): QTOP 2019, LNCS 11413, pp. 36–48, 2019.
https://doi.org/10.1007/978-3-030-14082-3_4

**Four Principles.** The discussion is organized around four key principles.

1. *Experimenters should be aware of unusual aspects of quantum computation.*
2. *Conclusions must be supported by the data.*
3. *Experiments should be reproducible by others.*
4. *Application benchmarks should be relevant to the concerns of the practitioner.*

Although these principles are not controversial, questions arise as to how best to meet them in experimental work on quantum platforms. The following sections present guidelines (marked with **(Gn)**) that align with these principles, for choosing experimental procedures and metrics, and for reporting results.

## 2   Unusual Aspects of Quantum Computation

Three unusual properties of quantum computation with implications for empirical work are discussed here. First, the technology is developing at an extremely rapid pace. Second, quantum computation takes place within a classical framework. Third, while open system quantum computing is always probabilistic (in the sense that any given computation may fail due to control errors and environmental noise), quantum annealers are "natural optimizers," which tend to fail in specific ways that have specific remedies.

*Rapid Development of Technology.* In classical computing we distinguish between the abstract algorithm, which has general properties that are hard to make precise (such as worst case $O(2^n)$ time), and the concrete solver, which has specific properties that are hard to generalize (such as 23.45 ms wall clock time).

In quantum computing we make a similar distinction; however, the abstract algorithm is assumed to run on an ideal processor in a noise-free system, whereas the concrete solver is subject to imperfections in fabrication, calibration, and shielding from environmental noise. An empirical test can only reveal the net performance of the ideal quantum algorithm, as reduced by limitations of the physical system.

Because of rapid improvements in the technology, empirical results about quantum performance tend to be somewhat ephemeral. Denchev et al. [10] report that in one experiment, a D-Wave 2X QPU solved a problem 10,000 times faster than their predictions based on an earlier D-Wave Two system, due to better noise suppression and colder refrigeration. They remark: "previous attempts to extrapolate the D-Wave runtimes ... will turn out to be of limited use in forecasting performance of future devices. For this reason, the current study focuses on runtime ratios that were actually measured ... rather than on extrapolations of asymptotic behavior which may not be relevant once we have devices which can attempt larger problems."

In these early days of quantum computing, hardware improvements that yield runtime speedups of a thousandfold and more have been observed. This trend is expected to continue for the foreseeable future, as qubit counts increase and better technologies and support tools continue to be developed.

(G1) Do not expect performance of one QPU to be predictive of performance of future systems. Do not rely too heavily on results from earlier-generation QPUs when developing new tests of quantum performance.

*Quantum Core in a Classical Framework.* In a D-Wave system, the quantum annealing algorithm runs on a QPU containing qubits and couplers that are driven by signals from a classical control system. The control system is part of a larger server-based system that queues and schedules jobs, invokes postprocessing utilities, bundles outputs for the user, and provides library support for related tasks.

This raises the question of which system components should "count" when measuring runtimes. Experimental guidelines in classical computation unequivocally recommend apples-to-apples comparisons—solver to solver, subroutine to subroutine, platform to platform. But where do we draw these lines when evaluating an algorithm that is implemented as a machine instruction in quantum hardware operating within a classical control framework?

A good rule of thumb is to consider classical and quantum solvers to be matched if they work on identically-formatted inputs and outputs. For example, *native* solvers for a D-Wave QPU work on Chimera-structured inputs and return spin values as outputs, whereas solvers for a factoring problem work on inputs and outputs that are integers. However, this rule does not completely settle the question: for example, QPU runtime can be defined as pure *anneal time*, preferred for direct comparison against classical solvers without including I/O overhead, or as *wall-clock time* (including programming, anneal, and readout) if I/O is considered part of the computation.

(G2) All system components used in the computation may be considered relevant when measuring runtimes. Note the use of *may be* not *must be*: the choice of component times to include in a study depends on the research question motivating the experiment. This choice should be made clear when reporting results.

Of course, the components used in the experiment may not match those used in applications or in future systems. For example, an experiment may minor-embed every input, while an application may only require one embedding for multiple inputs.

(G3) Report individual component times and invocation counts. This broadens applicability of the results by allowing readers to match components to their own use cases, and extends the lifetime of the results by allowing substitution of new component times as the technology evolves.

*A Natural Optimizer.* The Ising model and quadratic unconstrained binary optimization (QUBO) problems that are solved natively by the D-Wave QPU are *optimization* problems, defined in terms of an objective function that is used for scoring solution quality. The QPU can be used for a variety of other problem

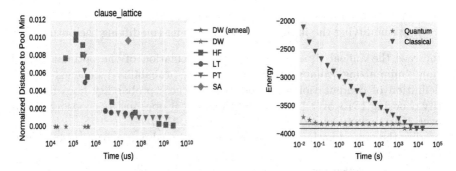

**Fig. 1.** Quantum annealing processors show fast convergence to good solutions. Left panel: the pink stars show QPU (labeled DW) performance measured as anneal time, and the red stars show QPU performance measured as wall-clock time. Also shown are results for Hamze-de Frietas-Selby (HF), a low-treewidth annealer (LT), parallel tempering (PT), and simulated annealing (SA). Right panel: typical patterns of convergence. (Color figure online)

types, including decision problems and sampling, but this section focuses on tests of optimization performance.

A common goal in optimization experiments is to study the tradeoff between computation time and solution quality, as illustrated in Fig. 1. The left panel shows solution quality versus computation time for a D-Wave 2000Q system (DW) and four native solvers, each tested under several parameter settings. Solution quality is measured as normalized mean distance to the minimum cost $C^*$ found by all solvers. Three versions of DW found $C^*$ in about .05 s anneal time (pink stars) and .5 s wall-clock time (red stars). The classical solvers show more gradual convergence patterns on this input: none have found $C^*$ within 1000 s, although HF is close.

The right panel shows a generic pattern of convergence for quantum annealers that has been observed in many contexts (e.g., [25,27]). Unassisted, the QPU finds good solutions very quickly, getting to the top gray line much faster than classical solvers (on large-enough inputs); but then it may struggle to move the sample minimum from "good" to "optimal," marked by the bottom gray line. All open system quantum computation is vulnerable to control errors and noise: in a quantum annealer that vulnerability is manifested as relatively slow convergence in the region between the two gray lines.

Note that if the underlying solution distribution is approximately Gaussian, then exponentially slow convergence time is guaranteed, since the sample minimum decreases only as the logarithm of sample size. In many cases, that time would be better spent applying tuning strategies aimed at moving the *location* of the underlying distribution closer to optimal.

Fortunately, a D-Wave system offers several tuning strategies when better-than-good solutions are needed, as listed below. The first four are parameters to the QMI, described in user documentation; the others are used during pre-processing (modifying the input) or post-processing (modifying the output).

1. Increase the anneal time. Note that the calculation of the optimal anneal time (minimizing [computation time/success probability]) depends on the definition of computation time: doubling success probability does not justify choosing $15\,\mu s$ over $10\,\mu s$ anneal time, but it does justify choosing $215\,\mu s$ over $210\,\mu s$ anneal-plus-readout time. See e.g., [19].
2. Modify the anneal path using anneal offsets, anneal pause, or reverse annealing [1,2,20].
3. Apply spin reversal transforms to "average out" small component biases.
4. Modify the problem Hamiltonian to increase the energy function gap between optimal and non-optimal solutions.
5. Modify the embedding to shrink maximum and mean chain lengths.
6. Adjust chain weights to find a balance between too-strong and too-weak.
7. Turn on the post-processing optimization utility.

(G4) Make use of QPU tuning techniques and system utilities to improve solution quality.

## 3  Conclusions Supported by the Data

A fundamental question in experimental algorithmics is how to generalize the results to draw trustworthy conclusions about performance. This section surveys problems that arise when studying solvers for NP-hard problems, and argues for caution when trying to extend conclusions beyond the bounds of the experiment.

*The Affinity Problem.* Suppose you have a quantum solver $Q$ in hand for the factoring problem: given an $n$-bit integer $N$, find a prime divisor $P < N$ or else report that $N$ is prime. For comparison you implement the classical Brute Force (BF) algorithm, which iterates $p = 2, 3, 4, \ldots \lfloor \sqrt{N} \rfloor$, testing whether $p$ divides $N$. If it finds a divisor, it stops and reports $P = p$; otherwise, upon reaching $\lfloor \sqrt{N} \rfloor + 1$, it reports that $N$ is prime.

Tests of BF on random $n$-bit integers reveal that on most inputs, BF stops within ten iterations.[2] Although the factoring problem is thought to be hard in the *worst case*, it is easy to solve in the *common case*. Do not expect $Q$ to show any significant speedup over BF in this experiment: there is no room for improvement when inputs are this easy. As well, fast runtimes for BF in this test are not representative of its enormous runtime on hard inputs (e.g., primes).

Next you implement a new solver BF2 that iterates downward from $p = \lfloor \sqrt{N} \rfloor$; upon finding a divisor $p^*$, it recurs with $N^* = p^*$, stopping when it determines that $N^*$ is prime. Note that although both BF and BF2 need one iteration in the best case and $\exp(n/2)$ iterations in the worst case, the specific inputs that invoke these costs are very different: easy inputs for one are hard for

---

[2] Because over 60% of integers are divisible by primes less than or equal to 11.

the other. How do you design a fair test using inputs that challenge both solvers equally?

This conundrum is fundamental to empirical work on NP-hard problems, for which no *provably* hard input sets—requiring exponential worst-case cost for all possible algorithms—are known. Furthermore, the No Free Lunch principle[3] [18,28] tells us that input hardness is relative, depending on how well the input structure aligns with the solver structure, and that any solver with fast solution times on one set of inputs will experience equally-slow solution times on another set of inputs. This gives rise to the *affinity problem*: The outcome of any empirical study of classical solvers using a finite input set can only reveal how well-matched the solvers are for that input set, and not how well they might perform in general.

(**G5**) Do not generalize classical (or quantum) performance results for one given input set, to draw conclusions about performance on input sets that have not been tested.

*How Does It Scale?* The emphasis on asymptotic analysis in algorithm research makes it tempting to extrapolate experimental results to predict performance on inputs (within a given set) that are larger than the ones actually tested.

Resist this temptation. In addition to guideline **G1** about extrapolating quantum performance results, extrapolating data from classical computational experiments to larger $n$ is also hazardous; Panny [24] illustrates what can go wrong.

Furthermore, standard guidelines in computer systems performance analysis deprecate extrapolation of results to larger parallel platforms using unvalidated models: Bailey [3] notes that this practice is "perplexing" because it is based on an *assumption* about how performance scales with system size, which is often the main question prompting the research. Bailey's observation applies to extrapolation of algorithmic performance data to larger $n$, when the underlying cost model is unknown.

Nor are methods for finding and fitting models to data trustworthy for extrapolation purposes. For example, regression-based techniques cannot distinguish whether a finite data set grows polynomially or exponentially in $n$: this is easy to see by noting that any finite set of $m$ data points can be fit *exactly* to a polynomial of degree $m - 1$, no matter how it was generated. McGeoch et al. [23] look at heuristic methods for finding leading terms of functions in algorithmic data sets, and conclude that every such method can be fooled.

(**G6**) Paraphrasing Hockney [12]: Only timings actually measured may be cited. Extrapolations and projections may not be employed for any reason.

That being said, it is useful and interesting to study how performance scales with respect to parameters of both inputs and solvers. This type of analysis involves methods of interpolation rather than extrapolation, sometimes using

---

[3] The No Free Lunch *theorem* is a formal result that applies to a certain class of algorithms; the No Free Lunch *principle* is a folklore rule that applies to heuristics. The theorem and the principle apply to classical but not to quantum solvers.

regression as a descriptive rather than inferential tool—i.e., fitting curves for sightline and comparison purposes, with no claims made about the true underlying model.

For example, the work of identifying and characterizing various input properties that drive solver performance—which sometimes have much greater impact than problem size—is often done using inputs generated according to parameters that vary within some finite range. For another example, papers about quantum speedup [11,14,16,17,19,26] study scaling performance with respect to problem sizes that are measured over a finite range.

# 4    Reproducibility of Experiments

Reproducibility means that an interested outsider can perform the same experiment using a similar test subject and test apparatus, and observe similar results as the original work, within statistical error. This of course is a fundamental tenant of sound empirical work in science.

At present, reproducibility of quantum performance results is hampered by limited availability, rapid evolution of the technology, and short machine lifetimes. However, indirect reproducibility experiments can be done by looking at quantum performance in relation to fixed set of classical solvers and test inputs, to reveal properties that persist across generations of machines.

In experimental algorithmics, the solver may be viewed as a test subject or as an apparatus used to study its abstract counterpart, the algorithm [21,22]. In this latter context, solvers typically offer parameters that can be set by the user at runtime: for example, one could combine BF and BF2 in a general Brute Force Solver (BFS) that allows the user to choose "increment" or "decrement," and to select an initial value for $p$. Because of the affinity problem, the observed performance of BFS on a given input is highly dependent on those parameters. That is, the "test apparatus" includes both the solver and its parameter values.

D-Wave platforms offer several runtime parameters as well. Some (e.g. anneal time and anneal path adjustments) modify the anneal process; others are part of the classical support framework, including problem transformation and embedding tools used before the anneal, and postprocessing utilities used after the anneal.

*Tuning and over-tuning.* A question much-discussed in the methodology literature (e.g., [3–5,15]) is: How much effort should be spent optimizing runtime parameters for best performance on a given input set? Of course, it is an ethical imperative to avoid comparing a well-tuned to an ill-tuned solver. And, it is important to evaluate well-tuned solvers to understand their true capabilities.

On the other hand, undocumented tuning efforts are not replicable, and may yield what Johnson [15] calls "a serious underestimate of the computing time needed to apply the algorithm to a new instance/class of instances." Even when well-documented, the use of pilot studies involving computationally-intensive "sweeps" of all possible parameter settings (which can take hours or days [5]), can yield irreproducible results when performance is reported for only the best parameterization, selected *post hoc* without assurance that the parameters actually mattered to the outcome (as opposed to being due to random variation in the experiment). Also, the sweep computation is not reproducible in application contexts if tuning the solver requires more time than solving a given set of input instances, as discussed in the next section.

**(G7)** Make a good-faith effort to tune all solvers equally well. Verify that results depending on optimal parameter settings are reproducible. Report the tuning procedure and the time spent tuning as part of the overhead cost of using the solver.

**(G8)** Quoting Johnson [15]: "If different parameter settings are to be used for different instances, the adjustment process must be well-defined and algorithmic, the adjustment algorithm must be described in the paper, and the time for the adjustment must be included in all reported running times."

Obviously, the reproducibility criterion fails if the report of test results does not contain sufficient details about procedures used in the experiment.

**(G9)** When possible, make available the inputs and solver code used in your experiments. When reporting results, include all information about the tests that materially affects the outcomes. This includes:

- Input generation methods and parameter settings. Number of inputs generated per sampling point.
- Quantum system model, qubit and coupler counts, and parameter settings. The methods used to transform and embed the inputs. Use of postprocessing utilities. System constants such as programming and readout times.
- Classical source code language, compiler optimization settings, and parameter settings used in the tests.
- Tuning policies and the amount of effort used in tuning. If relevant, time required to perform sweeps in a pilot study.
- System components that were included in the time measurements (for both quantum and classical systems). How runtimes were measured. Assessment of system load while measurements were being taken. The machine model and number of cores used for testing.

## 5   Relevance to the Practitioner

Benchmarking experiments involve comparisons of quantum solvers to a standard set of classical solvers and inputs that persist over time. The goal may be to track progress in quantum technology, or to address the practitioner's question: can it outperform my current solution method when applied to my use case?

In this context, "outperform" can have many meanings. Assuming $P \neq NP$, no algorithm for an NP-hard optimization problem can promise to deliver all three: (a) optimal solutions, (b) in polynomial time, (c) for every input. However, guarantees of two out of three are possible:

1. Some algorithms can find optimal solutions in polynomial time when run on restricted input sets, but may fail completely on general inputs. For example, an Ising model input can be solved in polynomial time if the input graph is planar.
2. Exact (also called complete) algorithms, such as branch-and-bound, guarantee to find optimal solutions to all inputs, but sometimes need exponential time.
3. An approximation algorithm can find solutions within some given ratio of optimal, in polynomial time, but cannot guarantee optimality.

It is safe to say that practitioners—a diverse group with a huge variety of applications and use cases in hand—are interested in all three dimensions of solver performance: *solution quality*, *computation time*, and *scope* (i.e., the range of inputs for which good performance is found). Heuristic solvers are also evaluated along these dimensions, although unlike algorithms, heuristics come with no theoretical guarantees, which makes empirical evaluation necessary. This section discusses guidelines for developing benchmarking tests that address the concerns of the practitioner.

*Variety of Inputs.* To address the third dimension of performance—solver scope—standard benchmark testbeds for algorithmic problems typically contain hundreds or thousands of instances, of the widest possible variety. The idea is to use input variety to (partially) overcome the challenges to generalization discussed in Sect. 3. Inputs may be obtained directly from real-world applications or from synthetic generators. Some generators are developed to mimic application inputs, adding controllable variety that might not otherwise be available; and some are developed from characterization studies aimed at exposing input properties that are most important to solver performance. An ideal testbed contains a good balance of inputs with properties that are challenging for a variety of solution approaches [22].

Standard testbeds for benchmarking quantum computers are not available in these early days while quantum performance mechanisms are rapidly evolving and little understood; however, it not too early to begin work towards that goal. One step in that direction is identification of test inputs that can differentiate solver performance.

(G10) Look at performance over a variety of input sets obtained from a variety of applications and from well-motivated generators.

*The Specialization Question.* Due to the affinity problem discussed earlier, different solvers and parameterizations show different performance on different inputs. A specialized solver tends to have many runtime parameters that can be tuned for extremely good performance on a given input set; but it has the drawback that extensive tuning may be necessary to avoid extremely slow performance on new inputs. A general solver tends to have few runtime parameters and is designed to be robust over a broad variety of input classes; that is, never extremely fast but never extremely slow.

Methodologists [5,15,21] argue that the practitioner prefers to use solvers that are general and robust, requiring little tuning overhead. As Barr et al. [5] put it: "the more specialized the problem structure, the greater the efficiencies that should result and the heavier the investigator's burden to demonstrate relevance and contribution." They also write, "if a heuristic performs well over a wide range of problems but fails on a specific type, then the authors should report, rather than hide, these results."

(G11) In benchmark studies of specialized and general solvers, focus on iden- tifying the most robust solver over a variety of input sets (best bad-case), rather than the fastest solver on any particular input set (best good-case).

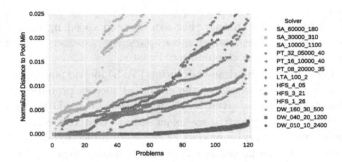

**Fig. 2.** A cactus plot captures solver performance over a range of inputs. Each solver was run for 500 ms wall-clock time on 120 inputs, 15 each from eight input classes.

*Variety of Metrics.* This guideline also argues for use of metrics and statistics that capture bad-case performance: reporting mean runtimes over all inputs is better than reporting median runtimes, and reporting statistics about data variability and extremal points is better yet. Hockney [12] suggests reporting the *specialty ratio,* defined as the ratio of maximum to minimum performance metrics observed over a range of test scenarios.

Graphical analysis does a better job than summary statistics at revealing performance from many perspectives. Figure 2 shows a *cactus plot* often used in heuristic benchmarking. Here, performance is measured as the best solution found within a fixed computation time; the solution quality metric is normalized distance to optimal $\delta$, as in Fig. 1. For each solver, the empirical cumulative distribution function (ECDF) of $\delta$ sampled on 120 inputs, is found by sorting $\delta$ in increasing order (note that vertically-matched points do not correspond to the same input instance). The red points correspond to a D-Wave 2000Q QPU tested under combinations of three parameters (annealing time, number of spin reversal transforms, and number of reads), labeled DW in the legend.

In terms of *worst case* performance—the height of the rightmost point for each solver—QPU solution quality was near $y = 0.0025$ in all three cases. In terms of *median* performance—the height of the ECDF curve on problem 60—we see that the QPU found minimum-cost solutions for at just over half the inputs tested. The *mean* performance of each solver corresponds to the area under each curve. Cactus plots are also useful for tests where solvers can fail (for example by timing-out without returning a solution), since unplotted failures create shorter curves.

(**G12**) Provide a full picture of performance by using a variety of performance metrics along different dimensions of performance, together with statistical methods that expose different features of the data.

*Exploiting Inside Knowledge.* Experiments that exploit information available in the laboratory setting can give valuable insights about performance while being efficient to carry out; however, they tend to yield results that cannot be reproduced by the practitioner. For example, to speed up data collection, a researcher might use hand-tuned solver parameters based on knowledge of parameters used to generate the inputs.

For another example, the time-to-solution (TTS) metric popular in quantum speedup studies is defined terms of the empirical success probability $\pi$ of finding an optimal solution in a given batch of solutions. Calculation of $\pi$ requires knowledge of the optimal cost; since exact solvers that can certify optimality are typically far too slow on input sizes of current interest, experiments measuring TTS incorporate the laboratory trick of generating inputs with "planted" solutions that are known by construction.

Johnson [15] deprecates the use of planted solutions and optimality metrics in studies of heuristic solvers. He argues that this approach is not relevant to application benchmarking because (1) such inputs do not usually resemble those found in practice, and (2) the practitioner does not expect to find optimal solutions—if optimality were required, an exact algorithm would be employed instead of a heuristic.

That being said, it is possible to imagine application scenarios in which the optimal cost is known, such as a sampling problem where optimal cost is calculated once and many optimal solutions are requested from the solver; also, NP-hard decision problems have the handy property that correctness (i.e. optimality) of any proposed solution can be verified efficiently. In these cases, an appropriate metric might be the mean number of samples needed until an optimal solution is found and verified.

(**G13**) Benchmarks should reflect a "no hints" policy, using test procedures and metrics that do not exploit information unavailable to the practitioner.

## 6    Final Remarks

Empirical performance analysis of quantum annealing processors is expanding to new research communities. The principles and guidelines discussed here aim to promote solid experimental work that is mindful of features of quantum technology, supports trustworthy conclusions, can be reproduced by others, and is relevant to practice.

# References

1. Reverse quantum annealing for local refinement of solutions. D-Wave Whitepaper, 14-1018A-A (2017)
2. Andriyash, E., et al.: Boosting integer factoring performance via quantum annealing offsets. 14-1002A-B (2016)
3. Bailey, D.H.: Misleading performance reporting in the supercomputing field. Sci. Program. **1**(2), 141–151 (1992)
4. Bailey, D.H.: 12 ways to fool the masses: Fast forward to 2011 (Powerpoint) crd.lbl.gov/~dhbailey (2011). An earlier version appeared as, 1991 "Twelve ways to fool the masses when giving performance results on parallel computers. Supercomputing Review **4**(8), 54–55
5. Barr, R.S., Golden, B.L., Kelly, J.P., Resende, M.G.C., Steward Jr., W.R.: Designing and reporting on computational experiments with heuristic methods. J. Heuristics **1**, 9–32 (1995)
6. Barr, R.S., Hickman, B.L.: Reporting computational experiments with parallel algorithms: issues, measures, and experts' opinions. ORSA J. Comput. **5**(1), 2–18 (1993)
7. Bartz-Beielstein, T., Chiarandini, M., Paquete, L., Preuss, M. (eds.): Experimental Methods for the Analysis of Optimization Algorithms. Springer, Heidelberg (2010). https://doi.org/10.1007/978-3-642-02538-9
8. Boixo, S., et al.: Characterizing quantum supremacy in near-term devices. Nat. Phys. **14**, 595–600 (2018)
9. Cohen, P.R.: Empirical Methods for Artificial Intelligence. MIT Press, Cambridge (1995)
10. Denchev, V.S., et al.: What is the computational value of finite range tunneling? Phys. Rev. X **5**, 031026 (2016)
11. Hen, I., Job, J., Albash, T., Rønnow, T.R., Troyer, M., Lidar, D.: Probing for quantum speedup in spin glass problems with planted solutions. arXiv:1502.01663 (2015)
12. Hockney, R.W.: The Science of Computer Benchmarking. SIAM, Philadelphia (1996)
13. Jain, R.: The Art of Computer Systems Performance Analysis. Wiley, Hoboken (1991)
14. Job, J., Lidar, D.A.: Test-driving 1000 qubits. arXiv:1706.07124 (2017)
15. Johnson, D.S.: A theoretician's guide to the experimental analysis of algorithms. In: Goldwasser, M.H., et al. (eds.) Data Structures, Near Neighbor Searches, and Methodology: Fifth and Sixth DIMACS Implementation Challenges. Discrete Mathematics and Theoretical Computer Science, AMS, vol. 59 (2002)
16. Katzgraber, H.G., Hamze, F., Zhu, Z., Ochoa, A.J., Munoz-Bauza, H.: Seeking quantum speedup through spin glasses: the good, the bad, and the ugly. Phys. Rev. X **5**, 031026 (2015)
17. King, A.D., Lanting, T., Harris, R.: Performance of a quantum annealer on range-limited constraint satisfaction problems. arXiv:1502.02089 (2015)

18. Macready, W.G., Wolpert, D.H.: What makes an optimization problem hard? Complexity **5**, 40–46 (1996)
19. Mandrà, S., Katzgraber, H.G.: A deceptive step towards quantum speedup detection. arXiv:1711.03168 (2018)
20. Marshall, J., Venturelli, D., Hen, I., Rieffel, E.G.: The power of pausing: advancing understanding of thermalization in experimental quantum annealers. arXiv:1810.05581 (2016)
21. McGeoch, C.C.: Toward an experimental method for algorithm simulation (feature article). INFORMS J. Comput. **8**(1), 1–15 (1995)
22. McGeoch, C.C.: A Guide to Experimental Algorithmics. Cambridge Press, Cambridge (2012)
23. McGeoch, C., Sanders, P., Fleischer, R., Cohen, P.R., Precup, D.: Using finite experiments to study asymptotic performance. In: Fleischer, R., Moret, B., Schmidt, E.M. (eds.) Experimental Algorithmics. LNCS, vol. 2547, pp. 93–126. Springer, Heidelberg (2002). https://doi.org/10.1007/3-540-36383-1_5
24. Panny, W.: Deletions in random binary search trees: a story of errors. J. Stat. Plan. Infer. **140**(8), 2335–2345 (2010)
25. Parekh, O., et al.: Benchmarking Adiabatic Quantum Optimization for Complex Network Analysis, volume SAND2015-3025. Sandia Report, April 2015
26. Rønnow, T.F., et al.: Defining and detecting quantum speedup. Science **345**(6195), 420–424 (2014)
27. Trummer, I., Koch, C.: Multiple query optimization on the D-Wave 2x adiabatic quantum computer. VLDB **9**, 648–659 (2016)
28. Wolpert, D.H., Macready, W.G.: No free lunch theorems for optimization. IEEE Trans. Evol. Comput. **1**, 67–82 (1997)

# Quantum Gate Algorithms

# Nash Embedding and Equilibrium in Pure Quantum States

Faisal Shah Khan[1(✉)] and Travis S. Humble[2]

[1] Center on Cyber-Physical Systems and Department of Mathematics,
Khalifa University, Abu Dhabi, United Arab Emirates
faisal.khan@ku.ac.ae
[2] Quantum Computing Institute, Oak Ridge National Lab, Oak Ridge, TN, USA
humblets@ornl.gov

**Abstract.** With respect to probabilistic mixtures of the strategies in non-cooperative games, quantum game theory provides guarantee of fixed-point stability, the so-called Nash equilibrium. This permits players to choose mixed quantum strategies that prepare mixed quantum states optimally under constraints. We show here that fixed-point stability of Nash equilibrium can also be guaranteed for pure quantum strategies via an application of the Nash embedding theorem, permitting players to prepare pure quantum states optimally under constraints.

**Keywords:** Nash embedding · Fixed-point stability · Nash equilibrium

## 1 Introduction

As quantum technologies increase in scale and complexity, constrained optimization of the underlying quantum processes will also increase in importance. A good example is the case of a quantum Internet as a network of quantum devices that process and relay quantum information using teleportation and entanglement swapping. Connectivity constraints arise naturally when optimizing classical network resources [1,2], and similar constraints manifest in the design and operation of quantum networks [3–5]. Similarly, quantum computers implementing fault-tolerant operations must optimize circuit design in order to minimize the decoherence due to the intrinsic physical noise in gate operation while also adhering to constraints in the circuit layout, scheduling, and parallelization. Finding optimal

This manuscript has been authored by UT-Battelle, LLC under Contract No. DE-AC05-00OR22725 with the U.S. Department of Energy. The United States Government retains and the publisher, by accepting the article for publication, acknowledges that the United States Government retains a non-exclusive, paid-up, irrevocable, worldwide license to publish or reproduce the published form of this manuscript, or allow others to do so, for United States Government purposes. The Department of Energy will provide public access to these results of federally sponsored research in accordance with the DOE Public Access Plan (http://energy.gov/downloads/doe-public-access-plan).

© Springer Nature Switzerland AG 2019
S. Feld and C. Linnhoff-Popien (Eds.): QTOP 2019, LNCS 11413, pp. 51–62, 2019.
https://doi.org/10.1007/978-3-030-14082-3_5

solutions to such constrained optimization problems are expected to maximize the efficiency and performance of quantum information technologies.

Non-cooperative game theory [6] studies optimization under constraints and can offer useful insights into the engineering of scalable quantum technologies such as optimal bounds on errors and their correction in quantum computations. Any quantum physical process modeled as a non-cooperative game describes a *quantum game* of the same type. The first instance of non-cooperative game-theoretic modeling of quantum physical processes appears to be the 1980 work of Blaquiere [7], where wave mechanics are considered as a two player, zero-sum (strictly competitive) differential game and a mini-max result is established for certain quantum physical aspects. The more recent and more sustained game-theoretic treatment of quantum physical processes was initiated in 1999 with the work of Meyer [8]. Meyer's work considered quantum computational and quantum algorithmic aspects of quantum physics as non-cooperative games.

The year 1999 also saw the publication of the paper [9] by Eisert et al. in which a quantum informational model for the informational component of two players games was considered. This consideration was in the same spirit as the consideration of randomizing in a game which produces the so-called *mixed* game played with mixed strategies. The quantum informational model of Eisert et al. produces a *quantized* game. The inspiration for considering extensions of the informational aspect of games to larger domains comes from John Nash's famous theorem [10] in economics which not only innovates the solution concept of non-cooperative games as an equilibrium problem but, for probabilistic extensions of finite non-cooperative games, also guarantees its existence.

The promise of a Nash equilibrium solution is a foundational concept for game theory as it may be used to guarantee the behavior for the non-cooperating players. The relative simplicity of the proof of Nash's theorem for the existence of an equilibrium in mixed strategies in conventional games relies entirely on Kakutani's fixed-point theorem [11]. For quantum games, Meyer established the existence of Nash equilibrium in mixed strategies, which are modeled as mixed quantum states, using Glicksberg's [12] extension of Kakutani's fixed point theorem to topological vector spaces.

In this contribution, we note that the Kakutani fixed-point theorem does not apply directly to quantum games played with pure quantum strategies. But, one can use Nash's embedding of compact Riemannian manifolds into Euclidean space [13] (Nash's other, mathematically more famous theorem) and, under appropriate conditions, indirectly apply the Kakutani fixed-point theorem to guarantee Nash equilibrium in pure quantum strategies. We begin with a mathematically formal discussion of non-cooperative game theory and fixed-points.

## 2    Non-cooperative Games and Nash Equilibrium

An $N$ player, non-cooperative game in normal form is a function $\Gamma$

$$\Gamma : \prod_{i=1}^{N} S_i \longrightarrow O, \tag{1}$$

with the additional feature of the notion of non-identical preferences over the elements of the set of *outcomes* $O$, for every "player" of the game. The preferences are a pre-ordering of the elements of $O$, that is, for $l, m, n \in O$

$$m \preceq m, \text{ and } l \preceq m \text{ and } m \preceq n \implies l \preceq n. \tag{2}$$

where the symbol $\preceq$ denotes "of less or equal preference". Preferences are typically quantified numerically for the ease of calculation of the payoffs. To this end, functions $\Gamma_i$ are introduced which act as the *payoff function* for each player $i$ and typically map elements of $O$ into the real numbers in a way that preserves the preferences of the players. That is, $\preceq$ is replaced with $\leq$ when analyzing the payoffs. The factor $S_i$ in the domain of $\Gamma$ is said to be the *strategy set* of player $i$, and a *play* of $\Gamma$ is an $n$-tuple of strategies, one per player, producing a payoff to each player in terms of his preferences over the elements of $O$ in the image of $\Gamma$.

A non-cooperative $N$-player quantum game in normal form arises from (1) when one introduces quantum physically relevant restrictions. We declare a pure (strategy) quantum game to be any unitary function

$$Q : \otimes_{i=1}^{N} \mathbb{C}P^{d_i} \longrightarrow \otimes_{i=1}^{N} \mathbb{C}P^{d_i} \tag{3}$$

where $\mathbb{C}P^{d_i}$ is the $d_i$-dimensional complex projective Hilbert space of pure quantum states. The latter are typically referred to as $d$-ary "quantum digits" or *qudits*. By analogy with mixed game extensions, where players' strategies are probability distributions over the elements of some set, the strategies of each player in a quantum game consist of quantum superpositions over the elements of a set of observable states in $\mathbb{C}P^{d_i}$. These strategic choices are then mapped by $Q$ into $\otimes_{i=1}^{N} \mathbb{C}P^{d_i}$, over the elements of which the players have non-identical preferences defined using the overlap of two qudits as the payoff functions.

The overlap of two qudits is a complex number in general. This is in contrast to the more standard practice in classical game theory of defining payoff functions that map into the real numbers. Indeed, in the current context of pure strategy quantum games, the expected value of an observable computed after quantum measurement can be taken as the payoff function mapping into the set of real numbers. However, as we will show later in detail, the non-linearity of the expected value of an observable fails to guarantee Nash equilibrium whereas the linearity of the overlap does not.

*Nash equilibrium* is a play of $\Gamma$ in which every player employs a strategy that is a best reply, with respects to his preferences over the outcomes, to the strategic choice of every other player. In other words, unilateral deviation from a Nash equilibrium by any one player in the form of a different choice of strategy will produce an outcome which is less preferred by that player than before. Following Nash, we say that a play $p'$ of $\Gamma$ *counters* another play $p$ if $\Gamma_i(p') \geq \Gamma_i(p)$ for all players $i$, against the $n - 1$ strategies of the other players in the countered $n$-tuple, and that a self-countering play is an (Nash) equilibrium.

Let $C_p$ denote the set of all the plays of $\Gamma$ that counter $p$. Denote $\prod_{i=1}^{N} S_i$ by $S$ for notational convenience, and note that $C_p \subset S$ and therefore $C_p \in 2^S$.

Further note that the game $\Gamma$ can be factored as

$$\Gamma : S \xrightarrow{\Gamma_C} 2^S \xrightarrow{E} O \qquad (4)$$

where to any play $p$ the map $\Gamma_C$ associates its countering set $C_p$ via the payoff functions $\Gamma_i$. The set-valued map $\Gamma_C$ may be viewed as a preprocessing stage where players seek out a self-countering play, and if one is found, it is mapped to its corresponding outcome in $O$ by the function $E$. The condition for the existence of a self-countering play, and therefore of a Nash equilibrium, is that $\Gamma_C$ have a fixed point, that is, an element $p^* \in S$ such that $p^* \in \Gamma_C(p^*) = C_{p^*}$.

In a general set-theoretic setting for non-cooperative games, the map $\Gamma_C$ may not have a fixed point. Hence, not all non-cooperative games will have a Nash equilibrium. However, according to Nash's theorem, when the $S_i$ are finite and the game is extended to its *mixed* version, that is, the version in which randomization via probability distributions is allowed over the elements of all the $S_i$, as well as over the elements of $O$, then $\Gamma_C$ has at least one fixed point and therefore at least one Nash equilibrium.

Formally, given a game $\Gamma$ with finite $S_i$ for all $i$, its mixed version is the product function

$$\Lambda : \prod_{i=1}^{N} \Delta(S_i) \longrightarrow \Delta(O) \qquad (5)$$

where $\Delta(S_i)$ is the set of probability distributions over the $i^{\text{th}}$ player's strategy set $S_i$, and the set $\Delta(O)$ is the set of probability distributions over the outcomes $O$. Payoffs are now calculated as *expected payoffs*, that is, weighted averages of the values of $\Gamma_i$, for each player $i$, with respect to probability distributions in $\Delta(O)$ that arise as the product of the plays of $\Lambda$. Denote the expected payoff to player $i$ by the function $\Lambda_i$. Also, note that $\Lambda$ restricts to $\Gamma$. In these games, at least one Nash equilibrium play is guaranteed to exist as a fixed point of $\Lambda$ via Kakutani's fixed-point theorem.

**Kakutani Fixed-Point Theorem:** *Let $S \subset \mathbb{R}^n$ be nonempty, compact, and convex, and let $F : S \to 2^S$ be an upper semi-continuous set-valued mapping such that $F(s)$ is non-empty, closed, and convex for all $s \in S$. Then there exists some $s^* \in S$ such that $s^* \in F(s^*)$.*

To see this, make $S = \prod_{i=1}^{N} \Delta(S_i)$. Then $S \subset \mathbb{R}^n$ and $S$ is non-empty, bounded, and closed because it is a finite product of finite non-empty sets. The set $S$ is also convex because its the convex hull of the elements of a finite set. Next, let $C_p$ be the set of all plays of $\Lambda$ that counter the play $p$. Then $C_p$ is non-empty, closed, and convex. Further, $C_p \subset S$ and therefore $C_p \in 2^S$. Since $\Lambda$ is a game, it factors according to (4)

$$\Lambda : S \xrightarrow{\Lambda_C} 2^S \xrightarrow{E_\Pi} \Delta(O) \qquad (6)$$

where the map $\Lambda_C$ associates a play to its countering set via the payoff functions $\Lambda_i$. Since $\Lambda_i$ are all continuous, $\Lambda_C$ is continuous. Further, $\Lambda_C(s)$ is non-empty, closed, and convex for all $s \in S$ (we will establish the convexity of $\Lambda_C(s)$ below; the remaining conditions are also straightforward to establish). Hence, Kakutani's theorem applies and there exists an $s^* \in S$ that counters itself, that is, $s^* \in \Lambda_C(s^*)$, and is therefore a Nash equilibrium. The function $E_\Pi$ simply maps $s^*$ to $\Delta(O)$ as the product probability distribution from which the Nash equilibrium expected payoff is computed for each player.

The convexity of the $\Lambda_C(s) = C_p$ is straight forward to show. Let $r, s \in C_p$. Then

$$\Lambda_i(r) \geq \Lambda_i(p) \quad \text{and} \quad \Lambda_i(s) \geq \Lambda_i(p) \tag{7}$$

for all $i$. Now let $0 \leq \mu \leq 1$ and consider the convex combination $\mu r + (1 - \mu)s$ which we will show to be in $C_p$. First note that $\mu r + (1 - \mu)s \in S$ because $S$ is the product of the convex sets $\Delta(S_i)$. Next, since the $\Lambda_i$ are all linear, and because of the inequalities in (7) and the restrictions on the values of $\mu$,

$$\Lambda_i(\mu r + (1 - \mu)s) = \mu\Lambda_i(r) + (1 - \mu)\Lambda_i(s) \geq \Lambda_i(p) \tag{8}$$

whereby $\mu r + (1 - \mu)s \in C_p$ and $C_p$ is convex.

Going back to the game $\Gamma$ in (1) defined in the general set-theoretic setting, Kakutani's theorem would apply to $\Gamma$ if the conditions are right, that is, whenever the image set of $\Gamma$ is pre-ordered and $\Gamma_i$ is both linear and preserves the pre-order.

Kakutani's fixed-point theorem can be generalized to include subsets $S$ of convex topological vector spaces, as was done by Glicksberg in [12]. Using Glicksberg's fixed-point theorem, one can show that Nash equilibrium exists in games where the strategy sets are infinite or possibly even uncountably infinite. As mentioned earlier in the Introduction, Meyer used Glicksberg's fixed-point theorem to establish the guarantee of Nash equilibrium in mixed quantum games. This is not surprising given that probabilistic mixtures form a convex structure, which is an essential ingredient for fixed-point theorems to hold on "flat" manifolds such as $\mathbb{R}^m$.

## 3    Pure Quantum Games and Nash Equilibrium

No fixed-point theorem guarantee for Nash equilibrium in pure quantum strategies is known to exist in the literature. This is surprising perhaps given the rich, albeit non-convex, mathematical structure of $\mathbb{C}P^n$. More precisely, $\mathbb{C}P^n$ has a compact Riemannian (Kahler in fact) manifold structure with positive sectional curvature with respect to the Fubini-Study metric [14]. We use the richness of the mathematical structure of $\mathbb{C}P^n$ here to produce a guarantee of Nash equilibrium in pure quantum strategies, under restrictive conditions, by invoking Nash's embedding theorem:

**Nash Embedding Theorem:** *For every compact Riemannian manifold $M$, there exists an isometric embedding of $M$ into $\mathbb{R}^m$ for a suitably large $m$.*

The Nash embedding theorem tells us that $\mathbb{C}P^n$ is diffeomorphic to its image under a length preserving map into $\mathbb{R}^m$. The homeomorphism underlying this diffeomorphism allows us to treat $\mathbb{C}P^n$ and its image inside $\mathbb{R}^m$ as topologically equivalent; hence, we can treat $\mathbb{C}P^n$ as a sub-manifold $S$ of $\mathbb{R}^m$ and look for a fixed-point guarantee for continuous set-valued functions $F$

$$\mathbb{C}P^n \xhookrightarrow{e} S \xrightarrow{F} 2^S \tag{9}$$

via Kakutani's fixed-point theorem, where $e$ is the Nash embedding. To this end, recall that the Kakutani fixed-point theorem requires $S$ to be compact and convex. Since $e$ is a homeomorphism and $\mathbb{C}P^n$ is compact, $S$ is compact in $\mathbb{R}^m$. However, $S$ is not necessarily convex as homeomorphisms do not preserve convexity in general. Note also that the linearity of Nash embedding $e$ would be insufficient to ensure convexity of $S$, for an element $q$ of $\mathbb{C}P^n$ is equivalent to all scalar multiples $\lambda q$, for $\lambda \neq 0$. This violates the definition of convexity, which requires the possibility that $\lambda = 0$.

But suppose for the moment that there exists a convex embedding $S$ of $\mathbb{C}P^n$ into $\mathbb{R}^m$. For the Kakutani fixed-point theorem to be applicable to $F$, one encases $S$ in a simplex $\Delta$, establishes Kakutani's theorem on $\Delta$ via barycentric subdivision and the Brouwer fixed-point theorem [15], and then constructs a *retract* function

$$R : \Delta \longrightarrow S \tag{10}$$

that fixes the points of $S$, that is,

$$R(s) = s \tag{11}$$

for all $s \in S$. One can visualize the action of $R$ as projecting $\Delta$ onto $S$, possibly in a geometrically convoluted way, and projecting the points of $S$ onto themselves. Next, one defines a set-valued function

$$F' : \Delta \longrightarrow 2^\Delta \tag{12}$$

as $F'(s) = F(R(s))$; this function is upper-semicontinuous, and its fixed points lie in $S$ and are therefore fixed points of $F$.

Therefore, whenever it is possible to construct $e$ so that $S$ is convex in some $R^m$, then as per the discussion in the preceding paragraph, Nash equilibrium in pure strategy quantum games is guaranteed. On the other hand, when $S$ is not convex, for example when $S = e\left(\mathbb{C}P^1\right) = \mathbb{S}^2 \subset \mathbb{R}^3$, then one can extend to its convex hull $\mathrm{Conv}(S)$ (taking into account the conditions outlined in the Caveats section below) via the convex hull operation, call it $\mathcal{C}$, and note that by Caratheodory's theorem [16] $\mathrm{Conv}(S)$ is compact. It is a well-established topological fact that a non-empty, convex compact subset of $\mathbb{R}^m$ is homeomorphic to the closed unit ball

$$\mathbb{B}^m = \left\{ (x_1, \ldots, x_m) \in \mathbb{R}^m : \sum_{j=1}^m x_j^2 \leq 1 \right\}. \tag{13}$$

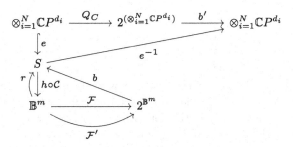

**Fig. 1.** Establishing the guarantee of Nash equilibrium in pure quantum strategies in a quantum game $Q$ using Nash's embedding theorem. The functions $b$ and $b'$ take a fixed-point from the power set back to the set.

Let $h$ be the homeomorphism $\text{Conv}(S) \cong \mathbb{B}^m$. The set-valued function $F$ in (9) now has factors

$$S \xrightarrow{\mathcal{C}} \text{conv}(S) \xrightarrow{h} \mathbb{B}^m \xrightarrow{\mathcal{F}} 2^{\mathbb{B}^m} \tag{14}$$

with continuous $\mathcal{F}$.

Next, set the payoff function $\mathcal{F}_i = Q_i$ to be any linear function with the property that the elements of its image can be pre-ordered so that (7) and (8) may be invoked. This produces convex countering sets $\mathcal{F}(h(\mathcal{C}(e(q))))$ of the image $e(q)$ of a play $q$ of $Q$ and guarantees a fixed point for $\mathcal{F}$. To ensure the existence of fixed-points of $\mathcal{F}$ over $S$, construct a retract

$$r : \mathbb{B}^m \longrightarrow S \tag{15}$$

and define the set-valued function

$$\mathcal{F}'(v) = \mathcal{F}(r(v)) \tag{16}$$

so that the fixed points of $\mathcal{F}'$ lie in $S$ and hence are the fixed points of $\mathcal{F}$. Mapping the fixed points back to $\mathbb{C}P^n$ via $e^{-1}$ gives a Nash equilibrium $q^*$ in $Q$.

This construction is captured in Fig. 1 for the relevant $n$. Note that when $S$ is convex, $\text{Conv}(S) = S$ and our construction restricts to an application of the Kakutani fixed-point theorem. Therefore, up to the caveats mentioned in the following section, our construction guarantees Nash equilibrium in general. Identifying the values of $m$ for which $\mathbb{C}P^n$ embeds as a convex subset into $R^m$ appears to be an open problem to the best of our knowledge.

## 4 Caveats

The Nash equilibrium guarantee due to the construction in Fig. 1 comes with the conditions that $S$ is non-empty and that $S$ does not equal the boundary of $\text{Conv}(S)$. We elaborate below.

Denote by $\partial$ the boundary of a non-empty set. Then,

$$h(\partial \text{Conv}(S)) \cong \partial h(\text{Conv}(S)) = \partial \mathbb{B}^m = \mathbb{S}^{m-1} \tag{17}$$

where

$$\mathbb{S}^{m-1} = \left\{ (x_1, \ldots, x_n) \in \mathbb{R}^n : \sum_{j=1}^{m} x_j^2 = 1 \right\} \tag{18}$$

is the $(m-1)$-dimensional sphere. Again, it is well-established that no retract exists from $\mathbb{B}^m$ to $\mathbb{S}^{m-1}$ [17] (the no-retract theorem). Therefore, if the Nash embedding $S = \partial \text{Conv}(S)$, then the construction in Fig. 1 will fail to guarantee Nash equilibrium. This is certainly true for $N = 1$ and $d_i = 1$ in the definition of a quantum game in (3), giving $\mathbb{C}P^1 = \mathbb{S}^2 = \partial \mathbb{B}^2$ (recall that $\mathbb{C}P^1$ $is$ $\mathbb{S}^2$ and hence the latter is the natural Nash embedding) as the single player's strategy set. Fortunately, since we are concerned only with multiplayer games, $N \geq 2$.

In general, the elements of any set are the extreme points of its convex hull, that is, points $x$ and $y$ in the convex hull for which $\lambda x + (1 - \lambda)y$ implies that either $\lambda = 0$ or $\lambda = 1$. Moreover, the set of extreme points generally forms a proper subset of the boundary of a convex set. Hence,

$$S \subset \partial \text{Conv}(S) \implies h(S) \subset \mathbb{S}^{m-1}. \tag{19}$$

Next, to see that it is possible to have a retract from $\mathbb{B}^m$ to $h(S) \subset \mathbb{S}^{m-1}$, note that every continuous function from $\mathbb{B}^m$ to $\mathbb{S}^{m-1}$ moves at least one point of the $(m-1)$-sphere by the no-retract theorem. The extreme case where all the points move is undesirable, but any other situation gives a subset $\mathcal{S}$ of fixed points of $\mathbb{S}^{m-1}$ over which a retract from $\mathbb{B}^m$ may be constructed. Setting $h(S) = \mathcal{S}$ will give a class of retracts to serve as the function $r$ in Fig. 1. For example, the function

$$f : (x_1, \ldots, x_m) \to \left( x_1, \ldots, x_{m-1}, \sqrt{1 - \sum_{j=1}^{m-1} x_j^2} \right) \tag{20}$$

has the "upper hemisphere" as retract.

## 5   An Application of Pure Strategy Quantum Games

As an application of Nash equilibrium in pure quantum strategies, consider the problem of preparing a $n$ qudit state. Due to decoherence errors, quantum states can deviate from some desired configuration. To model the state preparation and decoherence errors as a non-cooperative pure strategy quantum game, we introduce $N$ notional players so that each player prefers the prepared state to be closest to his desired configuration. A state preparing quantum mechanism $Q$ can now be viewed as a non-cooperative, pure strategy quantum game that maps a play of the game to a quantum superposition $q = \sum_{j=1}^{n} \alpha_j b_j \in \otimes_{i=1}^{n} \mathbb{C}P^{d_i}$ with $\sum |\alpha_j|^2 = 1$, and in which the payoffs are defined by the overlap of the prepared quantum state relative to the desired configuration of each player, or in other words, the inner-product of the two states. This linear payoff function is

$$Q_i(q) = \langle \psi_i, q \rangle \tag{21}$$

with $\psi_i \in \otimes_{i=1}^{N} \mathbb{C}P^{d_i}$ the preferred state of the $i$-th player. The fact that the complex numbers can be pre-ordered, combined with the linearity of $Q_i(q)$, allows one to invoke (7) and (8) to establish the convexity of $\mathcal{F}(h(\mathcal{C}(e(q)))) = C_{e(q)}$.

By an application of the Nash embedding theorem and the Kakutani fixed-point theorem, we conclude that this game has fixed-point guarantee of Nash equilibrium and that there is an optimal solution to the problem of preparing a pure quantum state under the constraint of decoherence. A similar game model can also be applied to parameter estimation problems, especially in the context of quantum logic gates [18,19].

In contrast, we may also consider the case of a quantum game in which the payoff is defined with respect to a physical observable, as is typically done in quantum game theory literature. A physical observable is represented by a linear Hermitian operator whose eigenstates define possible outcomes. We may define the expectation value of such an operator with respect to a prepared quantum state $q$ as the corresponding payoff. The payoff function $\bar{Q}_i$ calculates the expected value of $q$ to player $i$ via

$$\bar{Q}_i(q) = \sum_{j=1}^{n} e_j |\alpha_j|^2 \tag{22}$$

where the $e_j$ are real numbers that numerically reflect the preferences of player $i$ over observable states $b_j$ of $\otimes_{i=1}^{n} \mathbb{C}P^{d_i}$ and $|\alpha_j|^2$ is the probability with which $q$ measures as $b_j$. The payoff function $\bar{Q}_i$ is not linear in general. Hence, the convexity of $C_{e(q)}$ does not follow in general and neither does a fixed-point guarantee for the existence of Nash equilibrium. This result is consistent with the results based on the quantization schemes a la Eisert et al. which define payoffs via (22) and for which pure quantum strategy Nash equilibrium are known to not exist in general.

For details on how quantum games are realized experimentally, we refer the reader to [20]. For a complete game-theoretic model for designing two qubit quantum computational mechanisms at Nash equilibrium, we refer the reader to [21,22].

## 6   Future Applications of Quantum Games

State amplification quantum algorithms like the famous Grover's algorithm [23] may be viewed as non-cooperative games between a player and Nature. Consider the quantum state representing the item being searched for as the preferred configuration of the player with control over $x$ qubits, while Nature, with her $y$ qubits, prefers anything but this configuration, with $x + y = n$. A multiplayer model is also possible. If the players' payoffs are given linearly as in (21), then the algorithm has a Nash equilibrium solution. If the payoffs are computed via (22), then Nash equilibrium may not exist.

With respect to mixed strategies, a natural application of quantum games would be in the areas of quantum communication or stochastic quantum processes where a coalition of players (Alice and Bob) engage in a non-cooperative way with the eavesdropper (Eve) [24]. Alice and Bob want to amplify privacy of the communication whereas Eve does not, and in fact may want to decrease it. If the Alice and Bob coalition and Eve try to achieve their respective outcomes via random quantum processes, then Glicksberg's theorem will guarantee a Nash equilibrium. With this guarantee in place, mechanism design methods can be adopted to find an equilibrium.

An important class of quantum games would be those that study equilibrium behavior of the subset of generalized quantum measurements on finite dimensional systems known as local operations and classical communication (LOCC), a set that is both compact and convex [25]. Because LOCC is significant in many quantum information processes as the natural class of operations, and given its compact and convex structures, constructing a non-cooperative finite quantum game model for it would be a worthwhile effort.

Adiabatic quantum computing [26–28] can potentially benefit from the quantum game model. An adiabatic quantum computation starts with a system of $n$ qudits in its lowest energy state. A Hamiltonian $H_I$ is constructed that corresponds to this lowest energy state and another Hamiltonian $H_f$ is used to encode an objective function the solution of which is the minimum energy state of $H_f$. Finally, the actual adiabatic computation occurs as the interpolating Hamiltonian

$$H\left(s(t)\right) = s(t)H_I + \left(1 - s(t)\right)H_f \tag{23}$$

which is expected to adiabatically transform the lowest energy state of $H_I$ to that of $H_f$ as a function of the interpolating path $s(t)$ with respect to time $t$. For large enough $t$ values, adiabaticity holds; on the other hand, $t$ should be much smaller than its corresponding value in classical computational processes for $H(s(t))$ to constitute a worthwhile effort.

Note that under exponentiation, $H(s(t))$ corresponds to a time-dependent unitary map from $\otimes_{i=1}^{n}\mathbb{C}P^{d_i}$ to itself that we can view as a zero-sum, non-cooperative quantum game. This quantum game has a notional player I that prefers an element of $\otimes_{i=1}^{n}\mathbb{C}P^{d_i}$ that corresponds to the lowest energy state of $H_f$. Player II or Nature, prefers anything but this element. Players I and II can be given access to any division of qubits to manipulate respectively via pure quantum strategies. Again, if the payoffs to the players are linear, then a Nash equilibrium is guaranteed.

**Acknowledgments.** Faisal Shah Khan is indebted to Davide La Torre and Joel Lucero-Bryan for helpful discussion on the topic of fixed-point theorems.

# References

1. Albers, S., Eilts, S., Even-Dar, E., Mansour, Y., Roditty, L.: On Nash equilibria for a network creation game. In: Proceedings of the Seventeenth Annual ACM-SIAM Symposium on Discrete Algorithm, pp. 89–98 (2006)
2. Fabrikant, A., Luthra, A., Menva, E., Papdimitriou, C., Shenker, S.: On a network creation game. In: Proceedings of the Twenty-Second Annual Symposium on Principles of Distributed Computing, pp. 347–351 (2003)

3. Liu, B., Dai, H., Zhang, M.: Playing distributed two-party quantum games on quantum networks. Quant. Inf. Process. **16**, 290 (2017). https://doi.org/10.1007/s11128-017-1738-0

4. Scarpa, G.: Network games with quantum strategies. In: Sergienko, A., Pascazio, S., Villoresi, P. (eds.) QuantumComm 2009. LNICST, vol. 36, pp. 74–81. Springer, Heidelberg (2010). https://doi.org/10.1007/978-3-642-11731-2_10

5. Dasari, V., Sadlier, R.J., Prout, R., Williams, B.P., Humble, T.S.: Programmable multi-node quantum network design and simulation. In: SPIE Commercial+ Scientific Sensing and Imaging, pp. 98730B–98730B (2016)

6. Binmore, K.: Playing for Real. Oxford University Press, Oxford (2017)

7. Blaquiere, A.: Wave mechanics as a two-player game. In: Blaquiére, A., Fer, F., Marzollo, A. (eds.) Dynamical Systems and Microphysics, vol. 261, pp. 33–69. Springer, Vienna (1980). https://doi.org/10.1007/978-3-7091-4330-8_2

8. Meyer, D.: Quantum strategies. Phys. Rev. Lett. **82**, 1052–1055 (1999)

9. Eisert, J., Wilkens, M., Lewenstien, M.: Quantum games and quantum strategies. Phys. Rev. Lett. **83**, 3077 (1999)

10. Nash, J.: Equilibrium points in N-player games. Proc. Natl. Acad. Sci. USA **36**, 48–49 (1950)

11. Kakutani, S.: A generalization of Brouwer's fixed point theorem. Duke Math. J. **8**(3), 457–459 (1941)

12. Glicksberg, I.L.: A further generalization of the Kakutani fixed point theorem, with application to Nash equilibrium points. Proc. Am. Math. Soc. **3**, 170–174 (1952)

13. Nash, J.: The imbedding problem for Riemannian manifolds. Ann. Math. **63**(1), 20–63 (1956)

14. Bengtsson, I., Zyczkowski, K.: Geometry of Quantum States: An Introduction to Quantum Entanglement, 1st edn. Cambridge University Press, Cambridge (2007)

15. Browuer, L.: Ueber eineindeutige, stetige transformationen von Flächen in sich. Math. Ann. **69** (1910)

16. Caratheodory, C.: Math. Ann. **64**, 95 (1907). https://doi.org/10.1007/BF01449883

17. Kannai, Y.: An elementary proof of the no-retraction theorem. Am. Math. Mon. **88**(4), 264–268 (1981)

18. Teklu, B., Olivares, S., Paris, M.: Bayesian estimation of one-parameter qubit gates. J. Phys. B: Atom. Mol. Opt. Phys. **42**, 035502 (6pp) (2009)

19. Teklu, B., Genoni, M., Olivares, S., Paris, M.: Phase estimation in the presence of phase diffusion: the qubit case. Phys. Scr. **T140**, 014062 (3pp) (2010)

20. Khan, F.S., Solmeyer, N., Balu, R., Humble, T.S.: Quantum games: a review of the history, current state, and interpretation. Quant. Inf. Process. **17**(11), 42 pp. Article ID 309. arXiv:1803.07919 [quant-ph]

21. Khan, F.S., Phoenix, S.J.D.: Gaming the quantum. Quant. Inf. Comput. **13**(3–4), 231–244 (2013)

22. Khan, F.S., Phoenix, S.J.D.: Mini-maximizing two qubit quantum computations. Quant. Inf. Process. **12**(12), 3807–3819 (2013)

23. Grover, L.K.: A fast quantum mechanical algorithm for database search. In: Proceedings of the 28th Annual ACM Symposium on the Theory of Computing, p. 212, May 1996

24. Williams, B.P., Britt, K.A., Humble, T.S.: Tamper-indicating quantum seal. Phys. Rev. Appl. **5**, 014001 (2016)

25. Chitambar, E., Leung, D., Mančinska, L., Ozols, M., Winter, A.: Everything you always wanted to know about LOCC (but were afraid to ask). Commun. Math. Phys. **328**(1), 303–326 (2014)

26. Farhi, E., Goldstone, J., Gutmann, S., Sipser, M.: Quantum computation by adiabatic evolution (preprint). https://arxiv.org/abs/quant-ph/0001106
27. McGeoch, C.C.: Adiabatic Quantum Computation and Quantum Annealing. Morgan & Claypool Publishers Series. Morgan & Claypool Publishers, San Rafael (2014)
28. Lucas, A.: Ising formulations of many NP problems. Front. Phys. **2**, 5 (2014)

# A Quantum Algorithm for Minimising the Effective Graph Resistance upon Edge Addition

Finn de Ridder[1(✉)], Niels Neumann[2], Thijs Veugen[2,3], and Robert Kooij[4,5]

[1] Radboud University, Nijmegen, The Netherlands
f.deridder@alumnus.utwente.nl
[2] TNO, The Hague, The Netherlands
{niels.neumann,thijs.veugen}@tno.nl
[3] CWI, Amsterdam, The Netherlands
[4] iTrust Centre for Research in Cyber Security,
Singapore University of Technology and Design, Singapore, Singapore
robert_kooij@sutd.edu.sg
[5] Faculty of Electrical Engineering, Mathematics and Computer Science,
Delft University of Technology, Delft, The Netherlands

**Abstract.** In this work, we consider the following problem: given a graph, the addition of which single edge minimises the *effective graph resistance* of the resulting (or, *augmented*) graph. A graph's effective graph resistance is inversely proportional to its *robustness*, which means the graph augmentation problem is relevant to, in particular, applications involving the robustness and augmentation of complex networks. On a classical computer, the best known algorithm for a graph with $N$ vertices has time complexity $\mathcal{O}(N^5)$. We show that it is possible to do better: Dürr and Høyer's quantum algorithm solves the problem in time $\mathcal{O}(N^4)$. We conclude with a simulation of the algorithm and solve ten small instances of the graph augmentation problem on the *Quantum Inspire* quantum computing platform.

**Keywords:** Graph augmentation · Effective graph resistance ·
Dürr and Høyer's algorithm · Quantum Inspire

## 1 Introduction

Our society depends on the proper functioning of several infrastructural networks, whose main function is to distribute flows of critical resources. Representative examples include electrical networks, distributing power through transmission links, and water networks, distributing water through pipe lines. The edges

---

Parts of this work are heavily based on the contents of De Ridder's master's thesis, in particular, Sects. 2 and 4. Readers interested in a more extensive treatment of the subject matter discussed in each of these sections are referred to the thesis available at https://www.ru.nl/publish/pages/769526/z_finn_de_ridder.pdf.

© Springer Nature Switzerland AG 2019
S. Feld and C. Linnhoff-Popien (Eds.): QTOP 2019, LNCS 11413, pp. 63–73, 2019.
https://doi.org/10.1007/978-3-030-14082-3_6

of these networks resist the passage of electric current and water molecules, respectively, and their resistance is governed by well-established physical laws. The physical characteristics of resistance in individual edges play a crucial role in the *robustness* of the network as a whole [1,2,13]. For this reason, the *effective graph resistance*, a graph metric also referred to as *Kirchhoff index*, is often used as a robustness indicator for complex networks (see [7]). The relationship between this metric and robustness is negative: a lower effective graph resistance indicates a more robust network.

Now suppose we can add a single edge to the network. Then the question that naturally arises is: "The addition of which (single) edge maximises the robustness of the augmented graph?" As robustness and the effective graph resistance are inversely proportional, in terms of the latter the problem becomes the following:

**Definition 1 (Graph Augmentation Problem).** *Given a graph G, the addition of which* single *edge e minimises $R_{G+e}$, where $R_{G+e}$ denotes the effective graph resistance of G augmented with e.*

An exhaustive search for a solution takes time $\mathcal{O}(N^5)$, where $N$ is the number of vertices in $G$ [20]. The worst-case running time of an exhaustive search depends on (i) the number of non-existing or *candidate edges* in $G$, on (ii) the worst-case running time of the best known algorithm that outputs the effective graph resistance of the graph it takes as input, and on (iii) the time it takes to find the minimum value amongst the outcomes. The number of candidate edges is $\mathcal{O}(N^2)$, and the best known algorithms for computing the effective graph resistance take time $\mathcal{O}(N^3)$ to do so, finding the minimum is also $\mathcal{O}(N^2)$ and accordingly, an exhaustive search takes time $\mathcal{O}(N^5)$.

There are different ways to compute the effective graph resistance of a graph. Currently, the most efficient way is to first compute the eigenvalues of the Laplacian of $G$, and make use of the fact that the effective graph resistance is equal to the sum of the multiplicative inverses of the non-zero eigenvalues. That is, Van Mieghem [16] has proven that

$$R_G = N \sum_{k=2}^{N} \frac{1}{\lambda_k}, \tag{1}$$

where $N$ is the number of vertices in $G$ and $\lambda_2 \leq \ldots \leq \lambda_N$ are the non-zero eigenvalues of its Laplacian. The *symmetric QR algorithm* (see for example [9]) can be used to find the eigenvalues of the Laplacian and takes time $\mathcal{O}(N^3)$.

In Sect. 4, we show that Dürr and Høyer's quantum algorithm can solve our problem in time $\mathcal{O}(N^4)$. Note, however, that our application of their quantum algorithm does *not* exploit graph-theoretic short cuts to outperform an exhaustive search for a solution, but instead makes use of the intricacies of quantum computation to arrive at a speed-up.

A part of Dürr and Høyer's algorithm, the *oracle*, is application dependent. To solve the graph augmentation problem, the oracle will need to compute the effective graph resistance of certain graphs, and to do so efficiently it could use the symmetric QR algorithm—any algorithm for computing the effective graph resistance in time at most $\mathcal{O}(N^3)$ would work. Remember, however, that Dürr and Høyer's algorithm is a *quantum* algorithm, which means the oracle cannot "just" be a *classical* implementation of the symmetric QR algorithm—more details on the oracle will be given in Sect. 4.

Before we do so, we first discuss in Sect. 2 related works on the graph augmentation problem and applications of quantum algorithms to graph theory. Section 3 is a brief introduction to quantum computation and serves as a preparation for Sect. 4, which embodies our main contribution. Before we conclude in Sect. 6, we discuss in Sect. 5 our simulation of the algorithm presented in Sect. 4.

## 2    Related Work

Ghosh et al. [8] have studied the problem of assigning weights to the edges of a given graph $G$, such that $R_G$ is minimised. They also show that among all graphs on $N$ vertices, the graph for which the effective graph resistance is maximal is the path graph $P_N$, and that the complete graph $K_N$ has the smallest effective graph resistance. A theorem by Deng [4] gives the minimum effective graph resistance among all graphs $G$ on $N$ vertices that have $k$ bridges; a *bridge* of a not-necessarily-connected graph $G$ is an edge whose removal increases the number of components of $G$ [11]. Van Mieghem [16] has shown that, if $G$ is a connected graph on $N$ vertices, then

$$\frac{(N-1)^2}{\bar{d}} \leq R_G, \tag{2}$$

where $\bar{d}$ denotes the average degree of the $N$ vertices. The bound is tight, i.e., there exists a graph $G$ for which $R_G = (N-1)^2/\bar{d}$, namely the complete graph $K_N$. Finally, Wang et al. [20] have proposed a handful of heuristics for the graph augmentation problem and in [19], the problem is studied in the context of power transmission grids.

Before we briefly discuss related works on applications of quantum algorithms to graph theory, it is worthwhile to mention that, for a problem related to the graph augmentation problem, namely that of *maximising* the *algebraic connectivity* of the graph upon edge addition, a more efficient algorithm (than exhaustive search) has been found already: Kim's bisection algorithm [12] has a time complexity of $\mathcal{O}(N^3)$, which is better than an exhaustive search, which also for this problem, takes time $\mathcal{O}(N^5)$. Heuristics for the problem can be found in the work by Wang and Van Mieghem [17].

Perhaps the second best-known quantum algorithm, after Shor's algorithm, is Grover's algorithm [10]. In fact, Dürr and Høyer's algorithm is a generalisation of Grover's. Grover's algorithm searches an unsorted database, to find some particular record, and has a *query complexity* of $\mathcal{O}(\sqrt{n})$ where $n$ is the number of records in the database. A classical search, by contrast, requires exactly $n$ queries in the worst case. Grover's algorithm achieves a *quadratic speed-up*.

Based on Grover's algorithm, Dürr et al. [5] formulated a quantum algorithm that determines whether a graph on $N$ vertices is connected, solving the problem in time $\mathcal{O}(N\sqrt{N})$, up to logarithmic factors. A classical computer requires time of order $N^2$, in the worst case. They also gave efficient algorithms for some other graph-theoretic problems related to strong connectivity, minimum spanning trees and shortest paths.

## 3  Quantum Computation

The following is a very short introduction to quantum computation. For a more elaborate introduction to quantum computing we refer to the standard work on the subject by Nielsen and Chuang [14].

Classical computing relies on bits to perform operations. By manipulating bits, operations are performed and results obtained. Quantum computers in principle do the same with quantum bits, or *qubits*. The state of a qubit $|\psi\rangle$ can be described by a vector

$$|\psi\rangle = \alpha_0 \begin{pmatrix} 1 \\ 0 \end{pmatrix} + \alpha_1 \begin{pmatrix} 0 \\ 1 \end{pmatrix} = \alpha_0|0\rangle + \alpha_1|1\rangle,$$

with the last expression in the so-called *ket*-notation. The complex coefficients $\alpha_0$ and $\alpha_1$ are called *amplitudes* and are subject to $|\alpha_0|^2 + |\alpha_1|^2 = 1$, to make $|\psi\rangle$ a valid quantum state.

If both amplitudes are non-zero, $|\psi\rangle$ is said to be in a *superposition* of the states $|0\rangle$ and $|1\rangle$. The combined state of two qubits $|\psi_1\rangle$ and $|\psi_2\rangle$ is $|\psi_1\rangle \otimes |\psi_2\rangle$, where $\otimes$ denotes the *tensor product*. Often however, it is not possible to decompose a combined state into a tensor product. If such decomposition is not possible, the qubits are said to be *entangled*.

There is an important difference between classical logic gates and quantum gates: the latter need to be reversible. Accordingly, a quantum gate can be described by a unitary operator. Analogous to classical computing, problems can be solved on a quantum computer by applying the right quantum gates on the right qubits in the right order.

## 4  The Algorithm

As mentioned before, Dürr and Høyer's algorithm is based on a generalisation of the well-known quantum search algorithm by Grover [10]. The generalisation was formulated by Boyer et al. [3], two years after Grover made public his pioneering research.

Given an unsorted database $T$, Dürr and Høyer's algorithm is able to find the *index* of the smallest entry in $T$. As one might expect, $T$ will contain for each *candidate edge* $e$ of $G$ (i.e., each edge that is not in $G$ already) the effective graph resistance $R_{G+e}$ of the augmented graph $G+e$. Accordingly, the algorithm will return the index of the edge that we are looking for.

This brings up the question: "How does the algorithm acquire $T$?" After all, if $T$ would be an argument of the algorithm, there would be no need to use it: finding the minimum in an unsorted database, given the database, is trivial and takes time proportional to the number of entries in the database. The actual answer is a bit more involved, but at the same time exemplary of the difference between *quantum* algorithms and algorithms that run on conventional computers.

Conceptually, the algorithm constructs and *queries* $T$ at runtime. To that end, it is given a single argument known as the oracle: the oracle is a function $f$ which, given two valid indices $(i, i')$ of $T$, returns $i$, if $T[i] < T[i']$, and $i'$, otherwise. The algorithm queries the database through the oracle. First, it will choose uniformly at random an index $i_r$ and then "ask" the oracle whether there exists another index $i_1$ different from $i_r$ such that $T[i_1] < T[i_r]$, and if so, to return $i_1$ (otherwise the oracle will simply return $i_r$). In the unlikely event that the oracle returns $i_r$, the solution was found by a single guess and we are done. If instead a different index $i_1$ is returned, a new query is raised: "Does there exist an index $i_2$ such that $T[i_2] < T[i_1]$?" These steps are repeated until the solution is found. Bear in mind, however, that the preceding is only a simplified explanation of the algorithm, to give some intuition into its workings. The actual specification is given in the following section.

## 4.1  Specification

As before, let $T$ denote the unsorted database that we are searching through and $T[i]$ the entry at index $i$. The number of entries in $T$ equals $n$. Dürr and Høyer's algorithm is shown in Fig. 1.

The algorithm outputs $i_m$ with probability larger than $1/2$ after at least $\lceil 8\pi\sqrt{n} \rceil$ applications of the Grover iterate. Therefore, if the algorithm is run $c$ times, the probability that the minimum is among the results is at least $1 - (1/2)^c$, which converges rapidly to 1 as $c$ increases. For example, after running the algorithm $c = 8$ times, the minimum is part of the outcomes with a probability that is higher than 0.99.

## 4.2  Complexity Analysis

The loop described in step (3) should be interrupted only if, with probability at least $1/2$, $i_t$ equals $i_m$. Accordingly, the running time of the algorithm is largely dependent on how long it takes for the foregoing to hold.

We define $X$ to be the time it takes until $i_t := i_m$. By Markov's inequality,

$$P\{X < 2\mathbb{E}[X]\} > \frac{1}{2} \tag{4}$$

and therefore, if we interrupt the loop after running for time $2\mathbb{E}[X]$, we know $P\{i_t = i_m\}$ is at least $1/2$. The remainder of this section will be about computing the expectation of $X$.

**Input.** A quantum circuit $U_f$ that implements the oracle $f$.

**Output.** With probability $p > 1/2$, index $i_m$ such that $T[i_m]$ is the smallest entry in the database.

1. Define $\lambda$ to be $8/7$ (see [3]).
2. Choose threshold index $i_t \in \mathbb{Z}$, $0 \le i_t \le n - 1$, uniformly at random.
3. Repeat the following until the cumulative sum of the number of applied Grover iterates exceeds $\lfloor 8\pi\sqrt{n} \rfloor$. Afterwards, go to (4.).

    a. Use the generalisation of Grover's algorithm by Boyer et al. [3] to search for an index $i$ such that $T[i] < T[i_t]$. That is,

        i. Initialise $s = 1$.
        ii. Initialise two $n$-qubit registers as $\sum_{i=0}^{n-1} \frac{1}{\sqrt{n}} |i\rangle |i_t\rangle$.
        iii. Choose $l \in \mathbb{Z}$, $0 \le l < s$, uniformly at random.
        iv. Apply the Grover iterate $l$ times. The oracle $f$ is implemented by $U_f$ as follows:

$$U_f : |i\rangle |i_t\rangle \mapsto \begin{cases} -|i\rangle |i_t\rangle & \text{if } T[i] < T[i_t] \\ |i\rangle |i_t\rangle & \text{otherwise} \end{cases}. \tag{3}$$

        v. Measure the first register. Let $i'_t$ be the outcome. If
        - $T[i'_t] < T[i_t]$      let $i_t := i'_t$ and return, i.e. go to (a.);
        - otherwise      let $s := \min(\lambda s, \sqrt{n})$ and go to (ii.).

4. Return $i_t$.

**Fig. 1.** Dürr and Høyer's algorithm

In step (v) of Boyer et al.'s algorithm, $i_t$ is changed if $T[i'_t] < T[i_t]$, after which the algorithm returns. The algorithm is then run again, if the cumulative sum of the number of applied Grover iterates does not yet equal $l$. For that reason, each *possible* execution of Boyer et al.'s algorithm can be associated with a distinct $i_t$: the threshold index just before the algorithm is executed.

In the worst case, i.e. if the running time of Dürr and Høyer's algorithm is maximal, the initial threshold index $i_{t_0}$, which is chosen uniformly at random in step (2), is such that $T[i_{t_0}]$ is the largest value in $T$. In addition, every possible execution of Boyer et al.'s algorithm is realised: at the end of each execution, the threshold index $i_t$ is changed to $i'_t$ such that $T[i'_t]$ is the largest entry in the database smaller than $T[i_t]$. Necessarily, the number of times Boyer et al.'s algorithm is executed equals $n - 1$.

Fortunately for us, it is unlikely that all $n-1$ possible executions are required. Most will be skipped, automatically, simply because the result of Boyer et al.'s algorithm is an index $i'_t$ such that $T[i'_t]$ is not only smaller than $T[i_t]$, but also smaller than a handful of different entries each of which is also smaller than $T[i_t]$, but that were just unlucky and have not been chosen. More than that, it is impossible that, in the future, they are chosen.

The expectation of $X$ is the sum of the expected running times of each possible execution of Boyer et al.'s. That is, let $Y_k$ be a random variable defined as the running time of the $k$th possible execution. We have,

$$\mathbb{E}[X] = \mathbb{E}[Y_1] + \mathbb{E}[Y_2] + \cdots + \mathbb{E}[Y_{n-1}], \tag{5}$$

where $\mathbb{E}[Y_k]$ is $p_k L_k$: the product of the probability $p_k$ of execution $k$ taking place and the running time or length $L_k$ of execution $k$, given it takes place, respectively. Observe that the $k$th possible execution occurs directly after, and at no other point in time, threshold index $i_t$ becomes $x$ such that $T[x]$ is the $k$th largest entry in $T$. As a result, $p_k$ is equal to the probability that $i_t := x$.

In the same paper in which they present their algorithm, Dürr and Høyer prove by induction that $p_k = 1/r$ where $1 \le r \le n$ is the *ranking* of $T[x]$, which is 1 if the entry is minimal, and $n$ if it is maximal (see Lemma 1 in [6]). Also, note that $r = n - k + 1$. As a result, for example,

$$\mathbb{E}[Y_{n-1}] = \frac{1}{2}L_{n-1}, \tag{6}$$

because the $(n-1)$th execution, the last of all possible executions, occurs if and only if during an earlier execution $i_t := y$, such that the ranking $r$ of $T[y]$ is 2, which by Dürr and Høyer's lemma happens with probability $p = 1/r = 1/2$. We find that

$$\mathbb{E}[X] = \frac{1}{n}L_1 + \frac{1}{n-1}L_2 + \cdots + \frac{1}{2}L_{n-1}, \tag{7}$$

and are left with the assignment to find $L_k$.

The length $L_k$ of a single execution is the sum of the lengths of steps (ii) and (iv); steps (i), (iii), and (v) take a negligible amount of time and are, accordingly, neglected in the complexity analysis. The duration of (ii) is the same for all iterations: by convention, it is $\log_2 n$ [6]. More difficult is the determination of the duration of (iv), because it depends on $k$.

Boyer et al. show that the expected number of applications of the Grover iterate, until their algorithm finds the minimum, is at most $8m_0$, where

$$m_0 = \frac{n}{2\sqrt{(n-t)t}}. \tag{8}$$

In (8), $t$ is the number of solutions, which is known for each possible execution, and allows us to find $L_k$:

$$L_k = 8m_0 B + \log_2 n = \frac{4Bn}{\sqrt{(n-k)k}} + \log_2 n, \tag{9}$$

where $B$ is the complexity of a single query, i.e. the complexity of $U_f$. If $n$ is sufficiently large and $\sqrt{n+1} \approx \sqrt{n}$, it is possible to show that, consequently

$$\mathbb{E}[X] \le 4\pi B\sqrt{n} + \ln n \log_2 n. \tag{10}$$

After time at least twice the upper bound of (10), i.e. after time $8\pi B\sqrt{n} + 1.39 \log_2^2 n$ we should stop executing Boyer et al.'s algorithm and move on to step (4). That is, after at least $\lceil 8\pi\sqrt{n} \rceil$ applications of the Grover iterate, the probability that $i_t = i_m$ is at least $1/2$. Accordingly, Dürr and Høyer's algorithm runs in time $\mathcal{O}(B\sqrt{n})$ where $B$ is the time complexity of querying the oracle and $n$ the size of $T$. The size of $T$ will be equal to the number of edges not in $G$ and therefore always smaller than $N^2$.

What is $B$, the complexity of $U_f$? If we query the oracle as to whether $T[i] < T[i_t]$, it must first compute $T[i] = R_{G+e_i}$. As mentioned before, the *symmetric QR algorithm* can be used to compute the eigenvalues of the Laplacian $Q$ of $G + e_i$, which can then be used to compute $R_{G+e_i}$. Since any classical algorithm can be implemented efficiently on a quantum computer using only *Toffoli gates* [14], the time complexity $B$ of the oracle is equal to the complexity of the symmetric QR algorithm, which is $\mathcal{O}(N^3)$. Hence, the complexity of Dürr and Høyer's algorithm, applied to the problem of minimising the effective graph resistance of the augmented graph, becomes $\mathcal{O}(N^4)$.

## 5  Simulation

The algorithm presented in Sect. 4 has been implemented on the quantum computing platform *Quantum Inspire* [15], developed by QuTech. QuTech is an advanced research centre in the Netherlands where the Technical University of Delft and TNO collaborate together with industrial partners to build a quantum computer and a quantum internet. The platform allows for simulations (of quantum computers) of up to 31 qubits.

Programming using Quantum Inspire is possible both via a web interface and via a software development kit (SDK). The SDK makes it possible to construct hybrid quantum/classical algorithms and run quantum instructions from the command line.

An (arbitrary) labelling of the candidate edges of the graph is needed to implement the algorithm. The binary representation of these labels relates directly to the states of the different qubits in the implementation. For example, state $|011\rangle$ relates to label 3. With high probability, the algorithm outputs the candidate edge whose addition minimises the effective graph resistance of the augmented graph. The algorithm is iterative and hence has multiple passes and multiple rounds of measurements.

An example of an implementation of the algorithm for only a single Grover iteration is shown in Fig. 2, other iterations are similar. Note that the circuit is subdivided in three parts. The first part, *state preparation*, creates an equal superposition over all quantum states in the first register and sets the quantum state to $|i_t\rangle$ in the second register.

In the second part, a phase-oracle is used to flip the amplitude of specific quantum states. That is, the amplitude of quantum states of edges better than $i_t$ are flipped. A Toffoli-gate combined with two Hadamard-gates flips the amplitude of the $|111\rangle$-state. Using $X$-gates before and after the controls of the Toffoli-gate, amplitudes of other states are flipped, i.e. with $X$-gates the target has to be in the $|0\rangle$-state for the amplitude to be flipped, without $X$-gates in the $|1\rangle$-state. In the example in Fig. 2, the oracle flips the amplitudes of the states $|110\rangle$, $|010\rangle$, and $|101\rangle$. The $CNOT$-gate with adjoining $X$- and Hadamard-gates is for the first two states, flipping the amplitude of $|\cdot 10\rangle$. The amplitude of the $|101\rangle$-state is flipped using the Toffoli-gate with adjoining $X$- and Hadamard-gates. In this example, the edges labelled 2, 5, and 6 result in a lower effective graph resistance than edge $i_t = 3$.

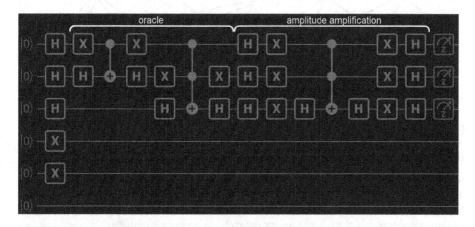

**Fig. 2.** The circuit implementation of the algorithm for $i_t = 3$. First, we have the state preparation part, afterwards a phase-oracle is applied using Hadamard-, $X$-, $CNOT$- and Toffoli-gates. Finally, amplitude amplification is applied and the first register is measured.

The last part of the algorithm is *amplitude amplification*: amplifying the amplitude of "good" states, while suppressing that of "bad" states. Note that the shown implementation is a non-optimised circuit-implementation, as for instance two Hadamard gates after each other do not change the quantum state.

We have tried the algorithm on ten small graphs, using only eight of the available 31 qubits. The ten graphs, labelled $G1, \ldots, G10$ as in Fig. 3, each consist of seven vertices and ten edges in total and are obtained from [18]. In the experiment we focused solely on the accuracy of the algorithm and not on the observed running times—after all, we are only simulating execution.

For each of the ten graphs we evaluated the algorithm a hundred times to test its accuracy and in each of the hundred runs, for each of the ten graphs, the best edge to add was found. We also found that, at least for these graphs, the upper bound on the number of iterations is a loose bound. For these graphs the upper bound would be $\left\lceil 8\pi\sqrt{11} \right\rceil = 84$, while logging information learned that for all graphs and for all hundred runs, after at most ten iterations the best edge was found already.

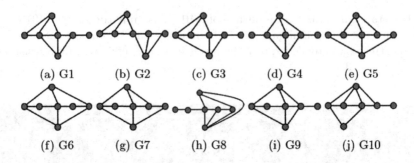

(a) G1      (b) G2      (c) G3      (d) G4      (e) G5

(f) G6      (g) G7      (h) G8      (i) G9      (j) G10

**Fig. 3.** The ten graphs $G1, \ldots, G10$ used to test the algorithm.

## 6   Conclusion

We have shown that the time complexity of Dürr and Høyer's algorithm for minimising the effective graph resistance is better than the time complexity of an exhaustive search. This means that there exists a quantum circuit that solves the optimisation problem and whose *depth*, compared to the depth of a classical circuit implementing an exhaustive search, increases significantly less fast as the number of vertices $N$ increases. As a result, it is likely that, at least for large instances, our algorithm running on a gate-based quantum computer will outperform an exhaustive search on a classical computer.

Our simulation of the algorithm shows that indeed with high probability, the best edge to add to the graph is found. In fact, in all runs, the solution was found. Our implementation is easily extended to also support larger graphs by increasing the number of qubits, adding extra controls to the controlled-NOT and Toffoli-gates, and by adding $X$-gates to the oracle.

Finally, it is very likely that there exist many other optimisation problems, for which also no clever algorithm has (yet) been formulated, and whose time complexity can be improved upon as well. Simply by taking advantage of the quadratic *reduction* of the time spent searching, provided by Dürr and Høyer's algorithm.

## References

1. Abbas, W., Egerstedt, M.: Robust graph topologies for networked systems. In: IFAC Proceedings, vol. 45, no. 26, pp. 85–90, September 2012. https://doi.org/10.3182/20120914-2-US-4030.00052, https://linkinghub.elsevier.com/retrieve/pii/S147466701534814X
2. Asztalos, A., Sreenivasan, S., Szymanski, B., Korniss, G.: Distributed flow optimization and cascading effects in weighted complex networks. Eur. Phys. J. B **85**(8) (2012). https://doi.org/10.1140/epjb/e2012-30122-3, http://www.springerlink.com/index/10.1140/epjb/e2012-30122-3
3. Boyer, M., Brassard, G., Høyer, P., Tapp, A.: Tight bounds on quantum searching. Fortschritte der Physik **46**(4–5), 493–505 (1998)
4. Deng, H.: On the minimum Kirchhoff index of graphs with a given number of cut-edges. MATCH Commun. Math. Comput. Chem. **63**, 171–180 (2010)

5. Dürr, C., Heiligman, M., Høyer, P., Mhalla, M.: Quantum query complexity of some graph problems. SIAM J. Comput. **35**(6), 1310–1328 (2006)
6. Dürr, C., Høyer, P.: A quantum algorithm for finding the minimum. arXiv:quant-ph/9607014, July 1996. http://arxiv.org/abs/quant-ph/9607014
7. Ellens, W., Spieksma, F., Van Mieghem, P., Jamakovic, A., Kooij, R.: Effective graph resistance. Linear Algebra Appl. **435**(10), 2491–2506 (2011). https://doi.org/10.1016/j.laa.2011.02.024, http://linkinghub.elsevier.com/retrieve/pii/S0024379511001443
8. Ghosh, A., Boyd, S., Saberi, A.: Minimizing effective resistance of a graph. SIAM Rev. **50**(1), 37–66 (2008). https://doi.org/10.1137/050645452, http://epubs.siam.org/doi/10.1137/050645452
9. Golub, G.H., Van Loan, C.F.: Matrix Computations. Johns Hopkins Studies in the Mathematical Sciences, 4th edn. The Johns Hopkins University Press, Baltimore (2013)
10. Grover, L.K.: A fast quantum mechanical algorithm for database search, pp. 212–219. ACM Press (1996). https://doi.org/10.1145/237814.237866, http://portal.acm.org/citation.cfm?doid=237814.237866
11. Harary, F.: Graph Theory. Perseus Books, Cambridge (2001)
12. Kim, Y.: Bisection algorithm of increasing algebraic connectivity by adding an edge. IEEE Trans. Autom. Control **55**(1), 170–174 (2010). https://doi.org/10.1109/TAC.2009.2033763, http://ieeexplore.ieee.org/document/5340588/
13. Koç, Y., Warnier, M., Mieghem, P.V., Kooij, R.E., Brazier, F.M.: The impact of the topology on cascading failures in a power grid model. Physica A: Stat. Mech. Appl. **402**, 169–179 (2014). https://doi.org/10.1016/j.physa.2014.01.056, https://linkinghub.elsevier.com/retrieve/pii/S0378437114000776
14. Nielsen, M.A., Chuang, I.L.: Quantum Computation and Quantum Information, 10th Anniversary edn. Cambridge University Press, Cambridge (2010)
15. QuTech: Quantum Inspire (2018). https://www.quantum-inspire.com/. Accessed 15 Nov 2018
16. Van Mieghem, P.: Graph Spectra for Complex Networks. Cambridge University Press, Cambridge (2011)
17. Wang, H., Van Mieghem, P.: Algebraic connectivity optimization via link addition. In: Proceedings of the 3rd International Conference on Bio-Inspired Models of Network, Information and Computing Sytems, BIONETICS 2008, pp. 22:1–22:8, ICST (Institute for Computer Sciences, Social-Informatics and Telecommunications Engineering), Brussels (2008). http://dl.acm.org/citation.cfm?id=1512504.1512532
18. Wang, X., Feng, L., Kooij, R.E., Marzo, J.L.: Inconsistencies among spectral robustness metrics. In: Proceedings of QSHINE 2018–14th EAI International Conference on Heterogeneous Networking for Quality, Reliability, Security and Robustness (2018)
19. Wang, X., Koç, Y., Kooij, R.E., Van Mieghem, P.: A network approach for power grid robustness against cascading failures. In: 2015 7th International Workshop on Reliable Networks Design and Modeling (RNDM), pp. 208–214. IEEE, Munich, October 2015. https://doi.org/10.1109/RNDM.2015.7325231, http://ieeexplore.ieee.org/document/7325231/
20. Wang, X., Pournaras, E., Kooij, R.E., Van Mieghem, P.: Improving robustness of complex networks via the effective graph resistance. Eur. Phys. J. B **87**(9) (2014). https://doi.org/10.1140/epjb/e2014-50276-0, http://link.springer.com/10.1140/epjb/e2014-50276-0

# Variational Quantum Factoring

Eric Anschuetz, Jonathan Olson, Alán Aspuru-Guzik, and Yudong Cao$^{(\boxtimes)}$

Zapata Computing Inc., Cambridge, MA, USA
`yudong@zapatacomputing.com`

**Abstract.** Integer factorization has been one of the cornerstone applications of the field of quantum computing since the discovery of an efficient algorithm for factoring by Peter Shor. Unfortunately, factoring via Shor's algorithm is well beyond the capabilities of today's noisy intermediate-scale quantum (NISQ) devices. In this work, we revisit the problem of factoring, developing an alternative to Shor's algorithm, which employs established techniques to map the factoring problem to the ground state of an Ising Hamiltonian. The proposed variational quantum factoring (VQF) algorithm starts by simplifying equations over Boolean variables in a preprocessing step to reduce the number of qubits needed for the Hamiltonian. Then, it seeks an approximate ground state of the resulting Ising Hamiltonian by training variational circuits using the quantum approximate optimization algorithm (QAOA). We benchmark the VQF algorithm on various instances of factoring and present numerical results on its performance.

**Keywords:** Integer factorization · Quantum computation · Discrete optimization

## 1 Introduction

Integer factorization is one of the first practically relevant problems that can be solved superpolynomially quicker on a quantum computer than on a classical computer [23]. Since its initial appearance, numerous follow-up studies have been carried out to optimize the implementation of Shor's algorithm from both algorithmic and experimental perspectives [1,5,11,17,21]. Improved constructions [1] have been proposed which, for an input number of $n$ bits, improve the circuit size from $3n$ qubits [19] to $2n + 3$ [1], and with nearest-neighbor interaction constraints [9]. It has also been pointed out that using iterative phase estimation [15], one can further reduce the qubit cost to $n + 1$, though the circuit needs to be adaptive in this case [17]. Various other implementations [12,24] of Shor's algorithm have been proposed such that only a subset of qubits need to be initialized in a computational basis state ("clean qubits").

Here we introduce an approach which we call *variational quantum factoring* (VQF). As with other hybrid classical/quantum algorithms such as the variational quantum eigensolver (VQE) [20] or the quantum autoencoder (QAE) [22],

© Springer Nature Switzerland AG 2019
S. Feld and C. Linnhoff-Popien (Eds.): QTOP 2019, LNCS 11413, pp. 74–85, 2019.
https://doi.org/10.1007/978-3-030-14082-3_7

classical preprocessing coupled with quantum state preparation and measurement are used to optimize a cost function. In particular, we employ the QAOA algorithm [6] and classical preprocessing for factoring. The VQF scheme has two main components: first, we map the factoring problem to an Ising Hamiltonian, using an automated program to find reductions in the number of required qubits whenever appropriate. Then, we train the QAOA ansatz[1] for the Hamiltonian using a combination of local and global optimization. We explore six instances of the factoring problem (namely, the factorings of 35, 77, 1207, 33667, 56153, and 291311) to demonstrate the effectiveness of our scheme in certain regimes as well as its robustness with respect to noise.

The remainder of the paper is organized as follows: Sect. 2 describes the mapping from a factoring problem to an Ising Hamiltonian, together with the simplification scheme that is used for reducing the number of qubits needed. Sect. 3 introduces QAOA and describes our method for training the ansatz. Sect. 4 presents our numerical results. We conclude in Sect. 5 with further discussion on future works.

## 2   Encoding Factoring into an Ising Hamiltonian

### 2.1   Factoring as Binary Optimization

It is known from previous work that factoring can be cast as the minimization of a cost function [2], which can then be encoded into the ground state of an Ising Hamiltonian [3,4]. To see this, consider the factoring of $m = p \cdot q$, each having binary representations $m = \sum_{k=0}^{n_m-1} 2^k m_k, p = \sum_{k=0}^{n_p-1} 2^k p_k, q = \sum_{k=0}^{n_q-1} 2^k q_k$ where $m_k \in \{0,1\}$ is the $k$th bit of $m$, $n_m$ is the number of bits of $m$, and similarly for $p$ and $q$. When $n_p$ and $n_q$ are unknown (as they are unknown *a priori* when only given a number $m$ to factor), one may assume without loss of generality [2] that $p \geq q$, $n_p = n_m$, and $n_q = \left\lceil \frac{n_m}{2} \right\rceil$ respectively[2]. By carrying out binary multiplication, the bits representing $m$, $p$, and $q$ must satisfy the following set of $n_c = n_p + n_q - 1 \in O(n_m)$ equations [2,4]:

$$0 = \sum_{j=0}^{i} q_j p_{i-j} + \sum_{j=0}^{i} z_{j,i} - m_i - \sum_{j=1}^{n_c-1} 2^j z_{i,i+j} \tag{1}$$

for all $0 \leq i < n_c$, where $z_{i,j} \in \{0,1\}$ represents the carry bit from bit position $i$ into bit position $j$. If we associate a clause $C_i$ over $\mathbb{Z}$ with each equation such that

$$C_i = \sum_{j=0}^{i} q_j p_{i-j} + \sum_{j=0}^{i} z_{j,i} - m_i - \sum_{j=1}^{n_c-1} 2^j z_{i,i+j}, \tag{2}$$

---

[1] The term *ansatz* means a trial solution of a specific form with some parameter(s) that need to be specified for the particular problem at hand.

[2] To lower the needed qubits for our numerical simulations, we assumed prior knowledge of $n_p$ and $n_q$.

then factoring can be represented as finding the assignment of binary variables $\{p_i\}$, $\{q_i\}$, and $\{z_{ij}\}$ which solves

$$0 = \sum_{i=0}^{n_c-1} C_i^2. \tag{3}$$

As a simple example, consider the factoring of $15 = 3 \times 5$. Long multiplication gives Table 1. Summing over columns then gives Eq. 2.

**Table 1.** A simple example explaining the derivation of Eq. 2. Here we factor the number 15, whose binary representation is 1111 (bottom row), into a product of two factors. Here $z_{i,j}$ denotes the carry bit from the $i$-th position to the $j$-th.

| | $p_2$ | $p_1$ | $p_0$ | |
|---|---|---|---|---|
| $\times$ | | $q_1$ | $q_0$ | |
| | $p_2 q_0$ | $p_1 q_0$ | $p_0 q_0$ | |
| $+\ p_2 q_1$ | $p_1 q_1$ | $p_0 q_1$ | | |
| $+\ z_{0,3}$ | $z_{0,2}$ | $z_{0,1}$ | $-2z_{0,1}$ | |
| $+\ z_{1,3}$ | $z_{1,2}$ | $-2z_{1,2}$ | $-4z_{0,2}$ | |
| $+\ z_{2,3}$ | $-2z_{2,3}$ | $-4z_{1,3}$ | $-8z_{0,3}$ | |
| 1 | 1 | 1 | 1. | |

## 2.2 Simplifying the Clauses

One method for simplifying clauses is to directly solve for a subset of the binary variables that are easy to solve for classically [3]. This reduction iterates through all clauses $C_i$ as given by Eq. (2) a constant number of times. In the following discussion, let $x, y, x_i \in \mathbb{F}_2$ be unknown binary variables and $a, b \in \mathbb{Z}^+$ positive constants. Along with some trivial relations, we apply the classical preprocessing rules[3]:

$$xy - 1 = 0 \implies x = y = 1,$$
$$x + y - 1 = 0 \implies xy = 0,$$
$$a - bx = 0 \implies x = 1,$$
$$\sum_i x_i = 0 \implies x_i = 0, \forall i \tag{4}$$
$$\sum_{i=1}^{a} x_i - a = 0 \implies x_i = 1, \forall i.$$

We also are able to truncate the summation of the final term in Eq. (2). This is done by noting that if $2^j$ is larger than the maximum attainable value of the

---

[3] We note that other simple relations exist that can be used for preprocessing—the simplified clauses for $m = 56153, 291311$ as used in our numerical simulations were given by [3] who utilized a different preprocessing scheme.

sum of the other terms, $z_{i,i+j}$ cannot be one; otherwise, the subtrahend would be larger than the minuend for all possible assignments of the other variables, and Eq. (1) would never be satisfied. This effectively limits the magnitude of Eq. (2) to be $O(n_m)$.

This classical preprocessing iterates through each of $O(n_c)$ terms in each of $n_c \in O(n_m)$ clauses $C_j$ (see Eqs. 1 and 2), yielding a classical computer runtime of $O(n_m^2)$. This is because $O(n_c) = O(n_m)$ from the identity $n_c = n_p + n_q - 1$, and $n_p \leq n_m$ and $n_q \leq \lceil \frac{n_m}{2} \rceil$ [2]. In practice, for most instances we have observed that the preprocessing program greatly reduces the number of (qu)bits needed for solving the problem, as is shown in Fig. 1.

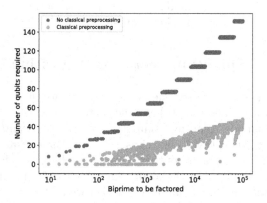

**Fig. 1.** This figure empirically demonstrates the reduction in qubit requirements after implementing the classical preprocessing procedure outlined in Sect. 2.2. After the classical preprocessing algorithm (orange), the number of qubits necessary for our algorithm empirically scales approximately as $O(n_m)$. In contrast, with no simplification (blue), VQF's qubit requirements scale as $O(n_m \log(n_m))$ asymptotically [2]. The discrete jump in the case of no preprocessing arises from the fact that many numbers of similar magnitudes can be represented with the same number of bits, and their factors require the same numbers of bits to represent also. The intermittent dropping in the orange point cloud arises due to factoring instances where the classical preprocessor is able to eliminate most of the variables from the system of simultaneous constraints. (Color figure online)

## 2.3 Constructing the Ising Hamiltonian

For each $i$ from 0 through $n_c - 1$, let $C_i'$ be $C_i$ after applying the classical preprocessing procedure outlined in Sect. 2.2. The solutions for the simplified equations $C_i' = 0$ then correspond to the minimization of the classical energy function

$$E = \sum_{i=0}^{n_c-1} C_i'^2, \tag{5}$$

which has a natural quantum representation as a *factoring Hamiltonian*

$$H = \sum_{i=0}^{n_c-1} \hat{C}_i^2. \tag{6}$$

Each $\hat{C}_i$ term is obtained by quantizing $p_i$, $q_i$, and $z_{j,i}$ in the clause $C_i'$ using the mapping $b_k \rightarrow \frac{1}{2}\left(1 - \sigma_{b,k}^z\right)$, where $b \in \{p, q, z\}$ and $k$ is its associated bit index. We have thus encoded an instance of factoring into the ground state of a 4-local Ising Hamiltonian. $H$ can also be represented in quadratic form by substituting each product $q_j p_{i-j}$ with a new binary variable $w_{i,j}$ and introducing additional constraints to the Hamiltonian [4]. This is necessary for implementation on quantum annealing devices with restricted pairwise coupling between qubits. However, in the gate model of quantum computation general methods for time evolution under $k$-local Hamiltonian are well known.

## 3  Variational Quantum Factoring Algorithm

The main component of our scheme is an approximate quantum ground state solver for the Hamiltonian in Eq. (6) as a means to approximately factor numbers on near-term gate model quantum computers. We use the *quantum approximate optimization algorithm* (QAOA), which is a hybrid classical/quantum algorithm for near-term quantum computers that approximately solves classical optimization problems [6]. The goal of the algorithm is to satisfy (i.e. find the simultaneous zeros of) the simplified clauses $C_i'$, which we cast as the minimization of a classical cost Hamiltonian $H_c$, and set to be identical to the Hamiltonian in Eq. (6) (i.e. $H_c = H$).

To prepare the (approximate) ground state we use an ansatz state

$$|\boldsymbol{\beta}, \boldsymbol{\gamma}\rangle = \prod_{i=1}^{s} \left(\exp\left(-i\beta_i H_a\right) \exp\left(-i\gamma_i H_c\right)\right) |+\rangle^{\otimes n}, \tag{7}$$

parametrized by angles $\boldsymbol{\beta}$ and $\boldsymbol{\gamma}$ over $n$ qubits, where $s$ is the number of layers of the QAOA algorithm. Here, $H_a$ is the *admixing Hamiltonian* $H_a = \sum_{i=1}^{n} \sigma_i^x$. For a fixed $s$, QAOA uses a classical optimizer to minimize the cost function

$$M(\boldsymbol{\beta}, \boldsymbol{\gamma}) = \langle \boldsymbol{\beta}, \boldsymbol{\gamma}| H_c |\boldsymbol{\beta}, \boldsymbol{\gamma}\rangle. \tag{8}$$

For $s \rightarrow \infty$, $M(\boldsymbol{\beta}, \boldsymbol{\gamma})$ is minimized when the fidelity between $|\boldsymbol{\beta}, \boldsymbol{\gamma}\rangle$ and the true ground state tends to 1. Generically for $s < \infty$, $|\arg\min(M(\boldsymbol{\beta}, \boldsymbol{\gamma}))\rangle$ may have exponentially small overlap with the true ground state. In our case, numerical evidence which will be discussed in Sect. 4 suggests that often letting $s \in O(n)$ suffices for large overlap with the ground state.

To optimize the QAOA parameters $\boldsymbol{\beta}$ and $\boldsymbol{\gamma}$, we employed a layer-by-layer iterative brute-force grid search over each pair $(\gamma_i, \beta_i)$, with the output fed into a BFGS global optimization algorithm [8]. The choices for grid sizes were motivated by a gradient bound given in [6]; more precisely, each dimension of the grid is $O\left(n_c^2 n^4\right)$. This can be shown as [6] gives a bound of $O\left(m^2 + mn\right)$ for each grid dimension for QAOA minimizing an objective function of $m$ clauses on $n$ variables. The setting in [6] is that each clause gives rise to a term in the Hamiltonian that has a norm at most 1. In our case, each clause $C_i'$ instead gives rise to a term in the Hamiltonian that has norm $\left\|\hat{C}_i^2\right\| \in O\left(n^2\right)$. Therefore, we take $m = n_c n^2$ and $n$ be the number of qubits, yielding a bound $O\left(n_c^2 n^4\right)$ for the gradient. To ensure that the optimum found by grid search differs from the true optimum by a constant, we therefore introduce a grid of size $O\left(n_c^2 n^4\right) \times O\left(n_c^2 n^4\right)$ based on the gradient bound. This ensures a polynomial scaling of the grid resolution. Numerically, training on coarser grids seemed sufficient (see Table 2).

**Table 2.** First column: Biprime numbers used in this study. Second column: the total number of qubits needed to perform VQF on the problem instance. Third column: among the qubits, the number of carry bits produced in the Ising Hamiltonian after simplifying the Boolean equations with rules described in (4). The observed difference between instances with carry bits versus without carry bits is shown in Fig. 2, along with Figs. 5 and 6. Fourth column: in the energy function (5), whether or not there exists a $p \leftrightarrow q$ symmetry. Such symmetry can be broken by two factors having different bit lengths. Fifth column: size of the grid used for the layer-by-layer brute-force search.

| Input number $m$ | Number of qubits $n$ | Number of carry bits | $p \leftrightarrow q$ symmetry | Grid size |
|---|---|---|---|---|
| $35 = 5 \times 7$ | 2 | 0 | ✓ | $6 \times 6$ |
| $77 = 7 \times 11$ | 6 | 3 | ✗ | $24 \times 24$ |
| $1207 = 17 \times 71$ | 8 | 5 | ✗ | $36 \times 36$ |
| $33667 = 131 \times 257$ | 3 | 1 | ✗ | $9 \times 9$ |
| $56153 = 233 \times 241$ | 4 | 0 | ✓ | $12 \times 12$ |
| $291311 = 523 \times 557$ | 6 | 0 | ✓ | $24 \times 24$ |

The remaining cost for finding the solution then comes from the global optimization procedure. In our numerical studies, the complexity scaling of performing BFGS optimization until convergence (to either a local or a global minimum) seemed independent of the problem size and depended linearly on the circuit depth (see Fig. 3). For a QAOA ansatz of depth $s$, this puts the total cost of performing VQF at $O\left(s^2 n_c^4 n^8\right)$ in the worst case, though numerically, this seems like a loose bound. We also note that there is no guarantee that this procedure always generates the globally optimal solution.

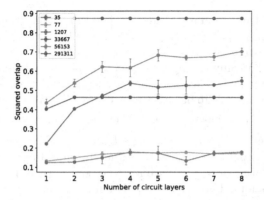

**Fig. 2.** The squared overlap of the optimized VQF state with the solution state manifold of $H_c$ for all problem instances considered. The squared overlap is also equal to the probability of successfully obtaining the solution upon measurement. Here, we fixed the error rate $\varepsilon = 10^{-3}$ and the number of samples $\nu = 10000$. We note the reduced depth scaling for $m = 77, 1207, 33667$ (see Sect. 4.1). The error bars each denote one standard deviation over three problem instances.

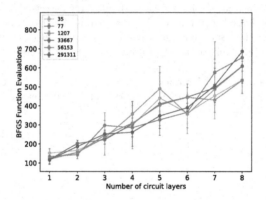

**Fig. 3.** The scaling of the number evaluations of Eq. (8) needed before the BFGS optimization converges. The scaling is approximately linear in the number of parameters, and is approximately independent of the problem size. The error bars each denote one standard deviation over three problem instances.

## 4    Numerical Simulations

### 4.1    Depth Scaling

We performed noisy simulation[4] of a number of instances of biprime factoring using the algorithm described above (see Sect. 4.2 for a description of

---

[4] The simulation of noisy quantum circuits was performed using QuTiP [13]. To access the data generated for all instances considered in this study, including those which produced Figs. 2, 3, 4, 5, and 6, please refer to our Github repository at https://github.com/zapatacomputing/VQFData.

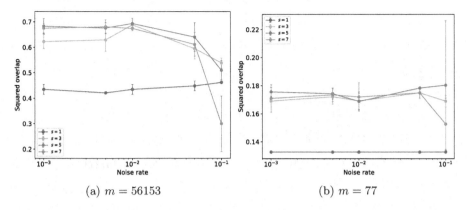

(a) $m = 56153$           (b) $m = 77$

**Fig. 4.** The dependence on factoring (a) $m = 56153$ and (b) $m = 77$ at various depths for different Pauli error noise rates. Below a certain error threshold, the success probability is approximately independent of the noise rate. The error bars each denote one standard deviation over three problem instances.

our noise model). Table 2 lists all of the instances used. With the technique described in Sect. 3, the success probability of finding the correct factors of $m = 35, 77, 1207, 33667, 56153, 291311$ as a function of the number of circuit layers $s$ is plotted in Fig. 2. The output distributions for representative numbers are plotted in Figs. 5 and 6. Here, "squared overlap" refers to the squared overlap of the output VQF state with the solution state manifold of $H_c$—that is, the squared overlap with states with the correct assignments of all $p_i$ and $q_i$ but not necessarily of all the carry bits $z_{ij}$, which are not bits of the desired factors $p$ and $q$.

For $m = 35, 56153, 291311$, after $O(n)$ circuit layers, the success probability plateaus to a large fraction. As factoring is efficient to check, one can then sample from the optimized VQF ansatz and check samples until correct factors of $m$ are found. However, the algorithm does not scale as well with the circuit depth for $m = 77, 1207, 33667$. This is the case even though the $m = 77, 33667$ problem instances have the same number or fewer qubits required than the $m = 56153, 291311$ problem instances. Further insight is needed to explain this discrepancy, though we do notice that unlike $m = 35, 56153, 291311$, these instances lack $p \leftrightarrow q$ symmetry and contain carry bits in their classical energy functions (5) (see Table 2).

### 4.2 Noise Scaling

An obvious concern for the scalability of the algorithm is the effect of noise on the performance of VQF. To explore this empirically, we considered a Pauli channel error model; that is, after every unitary (and after the preparation of $|+\rangle^{\otimes n}$) in Eq. (7), we implemented the noise channel $\rho \mapsto (1 - n\varepsilon) \rho + \frac{\varepsilon}{3} \sum_{j=1}^{n} \sum_{i=1}^{3} \sigma_j^{(i)} \rho \sigma_j^{(i)}$,

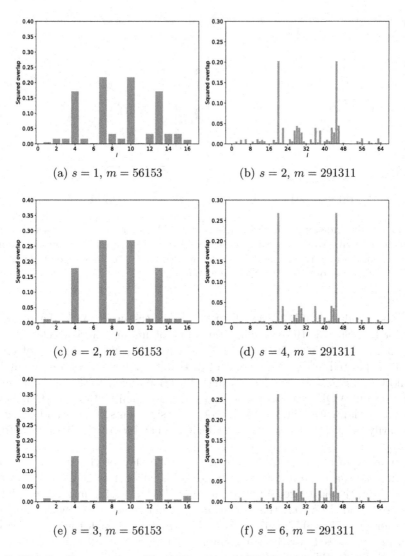

**Fig. 5.** Distributions corresponding to the output of the presented factoring algorithm for various circuit depths. $i$ labels computational basis states in lexicographic order. The two modes of each diagram correspond to the computational basis states yielding the correct $p$ and $q$; there are two modes due to the $p \leftrightarrow q$ symmetry of the problem. Here, we fixed the error rate $\varepsilon = 10^{-3}$ and the number of samples $\nu = 10000$.

where $\varepsilon$ is the single qubit error rate. Included in the simulation is sampling noise with $\nu = 10000$ samples when estimating the cost function $M(\boldsymbol{\beta}, \boldsymbol{\gamma})$. We plot the dependence of two VQF instances on the noise rate in Fig. 4, and note that VQF is weakly dependent on the noise rate below a certain error threshold.

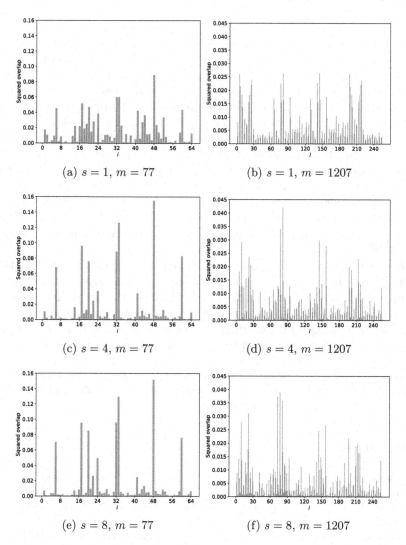

**Fig. 6.** Distributions corresponding to the output of the presented factoring algorithm for various circuit depths. Here $i$ labels computational basis states in lexicographic order. The modes of the high depth distributions are the correct ground states. We notice worse performance than $m = 56153, 291311$ (see Fig. 5 and Sect. 4.1). Here, we fixed the error rate $\varepsilon = 10^{-3}$ and the number of samples $\nu = 10000$.

# 5    Discussion and Conclusion

The ability to efficiently solve integer factorization has significant implications for public-key cryptography. In particular, encryption schemes based on abelian groups such as RSA and elliptic curves can be compromised if efficient factorization were feasible. However, an implementation of Shor's algorithm for factoring cryptographically relevant integers would require thousands of *error-corrected* qubits [14]. This is far too many for noisy intermediate-scale quantum devices that are available in the near-term, rendering the potential of quantum computers to compromise modern cryptosystems with Shor's algorithm a distant reality. Hybrid approximate classical/quantum methods that utilize classical pre- and post-processing techniques, like the proposed VQF approach, may be more amenable to factoring on a quantum computer in the next decade.

Although we show that it is in principle possible to factor using VQF, as with most heuristic algorithms, it remains to be seen whether it is capable of scaling asymptotically under realistic constraints posed by imperfect optimization methods and noise on quantum devices. We are currently in the process of examining more detailed analytical and empirical arguments to better determine the potential scalability of the protocol under realistic NISQ conditions. We look forward to working with our collaborators on experimental implementations on current NISQ devices.

In conclusion, the VQF approach discussed here presents many stimulating challenges for the community. QAOA, the optimization algorithm employed in our approach, has been studied by several groups in order to understand its effectiveness in several situations [6,7,10,16,18]. VQF inherits both the power and limitations of QAOA, and therefore many more numerical and analytical studies are needed to understand the power of VQF in the near future.

**Acknowledgments.** We would like to acknowledge the Zapata Computing scientific team, including Peter Johnson, Jhonathan Romero, Borja Peropadre, and Hannah Sim for their insightful and inspiring comments.

# References

1. Beauregard, S.: Circuit for Shor's algorithm using 2n+3 qubits. Quant. Inf. Comput. **3**(2), 175–185 (2003)
2. Burges, C.J.C.: Factoring as Optimization. Microsoft Research, MSR-TR-200 (2002)
3. Dattani, N.S., Bryans, N.: Quantum factorization of 56153 with only 4 qubits, arXiv:1411.6758 [quant-ph] (2014)
4. Dridi, R., Alghassi, H.: Prime factorization using quantum annealing and computational algebraic geometry. Sci. Rep. **7**(1), 43048 (2017)
5. Ekerå, M.: Modifying Shor's algorithm to compute short discrete logarithms. IACR Cryptology ePrint Archive (2016)
6. Farhi, E., Goldstone, J., Gutmann, S.: A quantum approximate optimization algorithm, arXiv:1411.4028 [quant-ph] (2014)
7. Farhi, E., Harrow, A.W.: Quantum supremacy through the quantum approximate optimization algorithm, arXiv:1602.07674 [quant-ph] (2016)
8. Fletcher, R.: Practical Methods of Optimization, 2nd edn. Wiley, New York (2000)
9. Fowler, A.G., Devitt, S.J., Hollenberg, L.C.L.: Implementation of Shor's Algorithm on a Linear Nearest Neighbour Qubit Array. Quantum Information & Computation **4**, 237–251 (2004)

10. Fried, E.S., Sawaya, N.P.D., Cao, Y., Kivlichan, I.D., Romero, J., Aspuru-Guzik, A.: qTorch: The quantum tensor contraction handler, arXiv:1709.03636 [quant-ph] (2017)
11. Geller, M.R., Zhou, Z.: Factoring 51 and 85 with 8 qubits. Sci. Rep. **3**(3), 1–5 (2013)
12. Gidney, C.: Factoring with n+2 clean qubits and n−1 dirty qubits, arXiv:1706.07884 [quant-ph] (2017)
13. Johansson, J., Nation, P., Nori, F.: QuTiP: an open-source Python framework for the dynamics of open quantum systems. Comput. Phys. Commun. **183**(8), 1760–1772 (2012)
14. Jones, N.C., et al.: Layered architecture for quantum computing. Phys. Rev. X **2**(3), 1–27 (2012)
15. Kitaev, A.Y.: Quantum measurements and the Abelian stabilizer problem, arXiv:quant-ph/9511026 (1995)
16. Lin, C.Y.-Y., Zhu, Y.: Performance of QAOA on typical instances of constraint satisfaction problems with bounded degree, arXiv:1601.01744 [quant-ph] (2016)
17. Martín-López, E., Laing, A., Lawson, T., Alvarez, R., Zhou, X.Q., O'brien, J.L.: Experimental realization of Shor's quantum factoring algorithm using qubit recycling. Nat. Photonics **6**(11), 773–776 (2012)
18. Nannicini, G.: Performance of hybrid quantum/classical variational heuristics for combinatorial optimization, arXiv:1805.12037 [quant-ph] (2018)
19. Nielsen, M.A., Chuang, I.: Quantum Computation and Quantum Information, 10th Anniversary edn. Cambridge University Press, Cambridge (2010)
20. Peruzzo, A., et al.: A variational eigenvalue solver on a photonic quantum processor. Nat. Commun. **5**(May), 4213 (2014)
21. Politi, A., Matthews, J.C.F., O'Brien, J.L.: Shor's quantum factoring algorithm on a photonic chip. Science **325**(5945), 1221–1221 (2009)
22. Romero, J., Olson, J., Aspuru-Guzik, A.: Quantum autoencoders for efficient compression of quantum data. Quant. Sci. Technol. **2**(4), 045001 (2017)
23. Shor, P.W.: Polynomial-time algorithms for prime factorization and discrete logarithms on a quantum computer. SIAM Rev. **41**(2), 303–332 (1999)
24. Zalka, C.: Shor's algorithm with fewer (pure) qubits, arXiv:quant-ph/0601097 (2006)

# Function Maximization with Dynamic Quantum Search

Charles Moussa[1,2]([envelope]) [ORCID], Henri Calandra[3], and Travis S. Humble[2] [ORCID]

[1] TOTAL American Services Inc., Houston, TX, USA
[2] Oak Ridge National Laboratory, Oak Ridge, TN, USA
charles.moussa@outlook.fr
[3] TOTAL SA, Courbevoie, France

**Abstract.** Finding the maximum value of a function in a dynamic model plays an important role in many application settings, including discrete optimization in the presence of hard constraints. We present an iterative quantum algorithm for finding the maximum value of a function in which prior search results update the acceptable response. Our approach is based on quantum search and utilizes a dynamic oracle function to mark items in a specified input set. As a realization of function optimization, we verify the correctness of the algorithm using numerical simulations of quantum circuits for the Knapsack problem. Our simulations make use of an explicit oracle function based on arithmetic operations and a comparator subroutine, and we verify these implementations using numerical simulations up to 30 qubits.

**Keywords:** Maximization · Quantum search · Quantum optimization

## 1 Introduction

Finding the maximal value in a poorly characterized function is a challenging problem that arises in many contexts. For example, in numerical simulations of particle dynamics, it is often necessary to identify the strongest interactions across many different particle trajectories. Prior results in quantum computing for unstructured search have shown that a quadratic speed up is possible when searching a poorly characterized function, and in this work, we consider an application of quantum search to the case of finding a maximal function value.

---

T. S. Humble—This manuscript has been authored by UT-Battelle, LLC, under Contract No. DE-AC0500OR22725 with the U.S. Department of Energy. The United States Government retains and the publisher, by accepting the article for publication, acknowledges that the United States Government retains a non-exclusive, paid-up, irrevocable, world-wide license to publish or reproduce the published form of this manuscript, or allow others to do so, for the United States Government purposes. The Department of Energy will provide public access to these results of federally sponsored research in accordance with the DOE Public Access Plan.

S. Feld and C. Linnhoff-Popien (Eds.): QTOP 2019, LNCS 11413, pp. 86–95, 2019.
https://doi.org/10.1007/978-3-030-14082-3_8

Quantum search was proposed originally to identify a marked item by querying an unstructured database [6]. By using an oracular implementation of the database function, Grover proved that a quadratic speedup in unstructured search could be obtained relative to brute force search by using superposition states during the querying phase. These ideas have been applied to a wide variety of application contexts including function maximization and minimization. In particular, Durr and Hoyer have shown how to design an iterative version of quantum search to find the minimizing argument for an unspecified oracle [5]. Their approach uses an updated evaluation of the oracle based on prior measurement observations, namely, to identify the smallest observed value. More recently, Chakrabarty et al. have examined the problem of using a dynamic Grover search oracle for applications in recommendation systems [3]. In addition, Udrescu et al. have cast the application of quantum genetic algorithms into a variant of dynamic quantum search [9].

In this contribution, we implement the principles of dynamic quantum search for the case of function maximization. We use the Knapsack 0/1 problem as an illustrative example, which is a constraint problem that finds the largest weighted configuration for a set of possible items. We extend recent work from Udrescu et al. [9] by using quantum circuits for arithmetic operations. Our implementation uses arithmetic operations to build the dynamic oracle for quantum search function maximization [1]. We test this implementation with a quantum program simulated using the Atos Quantum Learning Machine, which enables verification of the accuracy of the algorithm and estimation of the resources needed for its realistic implementation.

The paper is organized as follows: Sect. 2 formulates the quantum algorithm for function maximization using an iterative variant of quantum search, while Sect. 3 presents the explicit implementation of these ideas using quantum arithmetic circuits and results from numerical simulation of those circuits. Finally, we explain how quantum search is used in this context and show how we simulated the example step by step in Sect. 4.

## 2    Quantum Maximization of Dynamic Oracle

We show how to realize quantum search for function maximization using a dynamic oracle. As an illustrative example, we consider a Knapsack 0/1 problem introduced in Udrescu et al. as quantum version of a binary genetic algorithm called the reduced quantum genetic algorithm (RQGA) [9]. In particular, the genetic algorithm underlying RQGA was reduced to Grover search, thus removing the need for crossover and mutation operations for creating and selecting new candidate solutions. Rather, the complete set of candidate solutions is realized as an initial superposition over all the possible binary strings, e.g., using a Hadamard transform to prepare a uniform superposition state. We extend these ideas by developing an implementation using dynamic quantum search to solve the Knapsack problem.

Our method begins with an initial uniform superposition state of the candidate solutions stored in an $n$-qubit register $q$, where the size is determined by the number of possible items. The initial superposition is evaluated with respect to an objective function noted as $f$. For our example, the function $f$ computes the fitness of the candidate solution. The computed value for the fitness is stored in a second quantum register denoted as $f$. The register of size $p$ stores this value using two's complement form, where the size is determined by the largest value of all items in the database. The fitness is computed explicitly using a unitary operator $U_f$ what applies to both quantum registers $q$ and $f$, i.e.,

$$(1/\sqrt{2^n}) \sum_{i=0}^{2^n-1} |i\rangle_q |0\rangle_f \rightarrow (1/\sqrt{2^n}) \sum_{i=0}^{2^n-1} |i\rangle_q |f(i)\rangle_f \tag{1}$$

After preparing a uniform superposition of the candidate solutions and associated fitness values, the amplitudes with a fitness value greater than a current maximum value are marked by an oracle operator. The oracle operator is implemented explicitly below using a comparator circuit that takes as input the computed fitness register and a classical threshold value. The amplitude of all marked items are subsequently amplified to increase the probability to measure a better candidate. In practice, we extend the definition of the fitness operator to exclude certain candidates in the superposition as being valid or invalid candidate. For example, when valid candidates must have a positive fitness value, then a single register element may be used to indicate mark values that are negative and therefore invalid. Theoretically, Grover's search requires $\approx 13.6\sqrt{M}$ steps to achieve an error rate of less than 0.5 with $M$ being the size of the list searched [1].

Given this general overview of the algorithm, we next describe how to apply these ideas in a specific example for the Knapsack problem. Consider a backpack of $n$ possible items, where each item has a value and a weight. We seek the set of items that maximize the total value of the backpack while also not exceeding a defined total weight. Udrescu et al. [9] have used the example of $n = 4$ items, with the following properties: Item 1 (7 kg, \$40), Item 2 (4 kg, \$100), Item 3 (2 kg, \$50), and Item 4 (3 kg, \$30). The maximum weight allowed is 10 kg. Table 1 summarizes the set of all possible solutions for this example, from which it is apparent that the optimal solution is the set of all items excluding Item 1.

We summarize the computational steps taken to solve this example using the dynamic quantum search method within the quantum circuit model. Several quantum registers are used to store and process the problem input data, and we consider all registers to be initialized in the $|0\rangle$ state (Fig. 1).

- $q$: an $n$-qubit register for storing candidate solutions. $n = 4$ in this example.
- $f$: a $p$-qubit register for storing the value of the fitness calculation in two's complement representation, where $p$ is sufficient to store the sum of all possible values. $p = 6$ is this example.
- $w$: a register for storing the total mass of a candidate backpack. $\dim(w) = 5$ using unsigned integer representation in this example.
- $g$: a register for storing intermediate computational states and numerical constants. $\dim(g) = 6$ for this example.

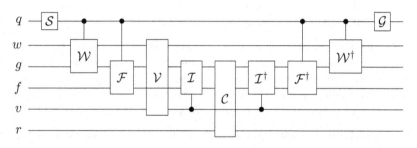

**Fig. 1.** This quantum circuit diagram represents the algorithm for function maximization using iterative Grover search with a dynamic oracle. We only represent with one Grover iteration for simplicity (more would require repeating the steps after creating the superposition). The quantum registers $q$, $w$, $g$, $v$, and $f$ define respectively storage space for the candidate solutions, computed weight, garbage, validity, and fitness values. Gates in the circuit are $\mathcal{S}$ for preparing the initial superposition of candidate states, $\mathcal{W}$ for computing candidate weight, $\mathcal{F}$ for computing candidate fitness value, $\mathcal{V}$ for computing validity (by loading the maximum weight in register $g$ and using a comparator with register $g$), $\mathcal{I}$ for inversing fitnesses of invalid candidates, $\mathcal{C}$ for comparing to the current maximum (after being loaded in register $g$), $\mathcal{G}$ represents the inversion about average operator. Measurement of the $q$ register should return a candidate that has a higher fitness than the current maximum loaded in $g$.

**Table 1.** Candidate solutions and values for Knapsack problem data.

| Candidate | Fitness | Weight | Validity |
|---|---|---|---|
| 0 0 0 0 | 0 | 0 | Valid |
| 0 0 0 1 | 30 | 3 | Valid |
| 0 0 1 0 | 50 | 2 | Valid |
| 0 0 1 1 | 80 | 5 | Valid |
| 0 1 0 0 | 100 | 4 | Valid |
| 0 1 0 1 | 130 | 7 | Valid |
| 0 1 1 0 | 150 | 6 | Valid |
| **0 1 1 1** | **180** | **9** | **Valid** |
| 1 0 0 0 | 40 | 7 | Valid |
| 1 0 0 1 | 70 | 10 | Valid |
| 1 0 1 0 | 90 | 9 | Valid |
| 1 0 1 1 | 120 | 12 | Invalid |
| 1 1 0 0 | 140 | 11 | Invalid |
| 1 1 0 1 | 170 | 14 | Invalid |
| 1 1 1 0 | 190 | 13 | Invalid |
| 1 1 1 1 | 220 | 16 | Invalid |

- $v$: a 1-qubit register to mark a candidate fitness as valid or invalid.
- $r$: a 1-qubit register used by the oracle.

Following the declaration and initialization of all registers, the candidate register $q$ is transformed under the $n$-fold Hadamard operator to prepare an uniform superposition of possible solution states. An initial fitness value threshold is selected by random number generation. We then apply a composite operator denoted as $o(x)$ that consists of the following stages:

1. Calculate the total mass, fitness, and validity of the candidates in registers $w$, $f$, and $v$.
2. Compare the fitness register $f$ against the current threshold value stored in $g$. If the fitness value is greater than the current threshold value, then set $r$ with the effect of marking the states, that is having the effect $|-\rangle \rightarrow (-1)^{o(x)}|-\rangle$ on the qubit oracle.
3. Uncompute the register $f$, $w$, and $v$.
4. We finally measure (after application of Grover iterations) the candidate solution register $q$ (and fitness register $f$ if we apply again the fitness computation circuit).

When computing the total mass and fitness, we store the weight and value of each item in the register $g$. Using a controlled adder, we perform addition with this register to store the sum of all weights and values for each candidate solution. As an illustration, consider the case that the fitness and mass are initially zero. We begin by adding the mass and value of the first item to the register. In our example problem, Item 1 has a mass 7 kg and value 4\$. We load the unsigned integer mass 7 (000111) into the workspace register $g$. The first qubit of the register $q$ acts as a control register for adding the mass of Item 1 in the register $m$ using controlled addition circuit. Similarly, we add the value 4 to the fitness register $f$ for those candidate solutions containing Item 1. This sequence of steps prepare the following quantum state (omitting workspace registers for clarity),

$$\sum_{i=0}^{7}|i\rangle_q|00000\rangle_w|000000\rangle_f + \sum_{i=8}^{15}|i\rangle_q|00111\rangle_w|000100\rangle_f \qquad (2)$$

The remaining items in the database are treated similarly.

As shown in Table 1, some candidate solutions have a total weight exceeding the constraint condition. These candidates are flagged by setting the register $v$ to 1, a step implemented by comparing the maximum weight constraint with the value of the register $w$. If the candidate weight exceeds the constraint, the register $v$ is flipped and we inverse their values to put them into their negative form so that when comparing to the current maximum, they will not be marked using the comparator quantum circuit operator. The negation operation is controlled by the validity qubit and implemented by complementing the fitness register $f$ and adding 1 to it as it is done in two's complement arithmetic.

$$\left(\sum_{i \in valid}|i\rangle_q|m(i)\rangle_w|f(i)\rangle_f|0\rangle + \sum_{i \in invalid}|i\rangle_q|m(i)\rangle_w|-f(i)\rangle_f|1\rangle_v\right)|001010\rangle_g \qquad (3)$$

At this stage, we wish to mark those candidate states that are valid and uncompute all values other than the candidate register. After preparing a superposition state representing valid possibilities, we use amplitude amplification to other registers as $|0\rangle$ and the qubit oracle as $|-\rangle$:

$$\sum_x |i\rangle_q (-1)^{o(i)} |-\rangle_r \tag{4}$$

We provide a pseudo-code representation for the steps before amplitude amplification in the Appendix. Grover iterations are finally used to change amplitudes. The number of required Grover iterations depends on the number size of the database as well as the number of actual solutions $M$ as

$$\left\lceil \frac{\pi}{4} \sqrt{\frac{N}{M}} \right\rceil \tag{5}$$

In our search for the maximizing solution, we do not know the number of solutions at each iteration in a maximization process with this quantum algorithm for finding the maximum. We therefore must use a version of quantum search that circumvents the need to estimate $M$. Boyer et al. [2] have provided methods for ensuring probabilistic success of the algorithm in the absence of knowledge about the number of items.

**Quantum Search for the Case $M$ is Unknown**

1. Initialize $m = 1$ and set a constant $\lambda = 6/5$.
2. Choose a random integer $j$ that is less than $m$.
3. Apply $j$ Grover routines over the superposition of candidates
4. In our case, apply again the part 1 of the oracle to have a fitness and the validity. We can also do this part with a classical function.
5. Measure to get an outcome candidate/fitness/validity.
6. If we get a valid solution and the fitness is greater than the current max, we stop and return them as new solution and new current max.
7. Else set $m$ to $\min(\lambda m, \sqrt{N})$ and repeat from step 2.

We may not have a solution in the case we already have the maximum of the problem but an appropriate time-out or early stopping can be set and the complexity would remain $\sqrt{N}$.

As an illustration, consider the case that the current threshold is 13. Candidate states in the superposition for our example problem will be marked if their associated weight is strictly greater than 13, as is the case for candidates 0110 and 0111. The resulting superposition becomes

$$\frac{1}{4} \left( \sum_{i \neq 0110, i \neq 0111} |i\rangle - |0110\rangle - |0111\rangle \right) \tag{6}$$

The result of the Grover search prepares the state

$$\frac{1}{8} \left( \sum_{i \neq 0110, i \neq 0111} |i\rangle \right) - \frac{5}{8} \left( |0110\rangle + |0111\rangle \right) \tag{7}$$

which yields acceptable candidate solutions with a probability $|5/8|^2 \approx 0.39$.

# 3   Quantum Circuit Simulation

We describe our implementation of these algorithm using integer arithmetic with unsigned integers presented in a binary representation and signed integers presented as two's complement. Let $a = (a_0, a_1, \ldots, a_{n-1})$ be the $n$-bit representation of the integer $A$, such that $A = \sum_i a_i 2^i \in [0, 2^n - 1]$. For signed integers, one bit is used for the sign and we can represent numbers $[-2^{n-1}, 2^{n-1} - 1]$. The most significant bit represents the sign, which is '+' for 0 and '−' for 1. Given two integers $a$ and $b$, bitwise addition is used to calculate the resultant $a + b$. For subtraction, recall that $a - b = (a' + b)'$ where $x'$ is the bit-wise complement of $x$ obtained by flipping all bits. Usually, an extra bit is required for addition but when not used, we actually do modular addition. To compare two numbers $a$ and $b$, one can do the subtraction and look at the last bit of the subtraction result. When it is 1, $a < b$.

The quantum ripple adder acts on inputs $a$ and $b$ by encoding those binary representations as the $|a, b\rangle$. The resulting output from in-place addition is the state $|a, a + b\rangle$. Note that we use another bit called high bit as the addition result may exceed $2^n$. In Ref. [8] a quantum circuit without ancilla qubit and with less quantum operations than previously proposed circuits was designed. This approach uses the Peres gates described in Fig. 2.

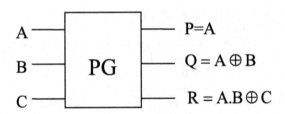

**Fig. 2.** Peres gate representation.

Cuccaro et al. previously described how the quantum ripple-adder may be adapted for other arithmetic tasks [4], and we make use of these implementations for our oracle definitions. For example, we define a subtraction method using the complement technique mentioned above. For modular addition, we consider the last significant bit of the number $b$ as the high bit, and apply the adder on the $n - 1$ first bits of $a$ and $b$, and apply a CNOT onto qubits representing the last significant bits. For a comparator, we use only the part of the adder where we compute the high bit and uncompute. If the high bit is 1, $a < b$. For a controlled adder, where given a control qubit we apply the adder if the control is $|1\rangle$, a quantum circuit was proposed in [7].

For our simulations, we use the Atos Quantum Learning Machine, a numerical simulation environment designed for verifying quantum software programs. It is dedicated to the development of quantum software, training and experimentation. We use the QLM as a programming platform and a high-performance quantum simulator, which features its own universal quantum assembly programming language (AQASM, Atos Quantum Assembly Language) and a high-level quantum hybrid language (PyQASM) embedded in the popular Python language. The QLM is capable of performing exact simulations of up to about 30 qubits using a vector-based representation of register states. The simulator manages the execution of the programmed sequence of quantum gates.

For our small example instance, we can find the optimal backpack in two Grover iterations (or two circuits of one Grover iteration). Some instances require more due to the probabilistic measurement outcomes, but a typical execution converges to valid solutions. It is possible to use the suboptimal solution as the threshold for the next instance of the algorithm in order to verify if better solutions are available. We just need a set of $n$ qubits to represent all possibilities while classically, this would be represented by many bit strings to form a population of candidates that will change over iterations like it is done in genetic algorithms.

## 4   Conclusion

In this contribution, we have shown how to implement function maximization using dynamic quantum search. Our approach has applied an explicit oracle for the quantum search algorithm and implemented a quantum algorithm for binary optimization of the Knapsack problem. Our implementation illustrates the idea of evaluating the fitness of all possible candidate solutions. However, our approach does require uncomputation of any ancillary register to ensure the correctness of the Grover iteration. While the added complexity is manageable, it would be advantageous to develop alternative search methods that act on subspaces of entangled registers. We anticipate that such methods would avoid the need for uncomputation of the ancillary registers and, in our example, permit more efficient steps to iteratively isolate the maximal fitness value.

**Acknowledgments.** CM and HC acknowledge support from Total and TSH acknowledges support from the U.S. Department of Energy, Office of Science, Early Career Research Program. Access to the Atos Quantum Learning Machine was provided by the Quantum Computing Institute at Oak Ridge National Laboratory.

# A    Pseudo-code for the Knapsack Example

---

**Algorithm 1.** Pseudo-code of the quantum circuit for the steps before amplitude amplification.

---

**Require:** Current maximum value, quantum circuit with all qubits prepared in $|0\rangle$.

$\triangleright$ Compute the weight $w_i$ of the candidates.

1: Apply Hadamard Transform on register $q$ to get all possible candidates in superposition.
2: Set register $r$ as $|-\rangle$.

3: Load $w_1$ in register $g$.
4: Control-add $w_1$ into register $w$ if the first qubit of register $q$ is 1.
5: Load $w_2$ in register $g$.
6: Control-add $w_2$ into register $w$ if the second qubit of register $q$ is 1.
7: Load $w_3$ in register $g$.
8: Control-add $w_3$ into register $w$ if the third qubit of register $q$ is 1.
9: Load $w_4$ in register $g$.
10: Control-add $w_4$ into register $w$ if the fourth qubit of register $q$ is 1.

$\triangleright$ Compute the fitness $f_i$ of the candidates.

11: Load $f_1$ in register $g$.
12: Control-add $f_1$ into register $f$ if the first qubit of register $q$ is 1.
13: Load $f_2$ in register $g$.
14: Control-add $f_2$ into register $f$ if the second qubit of register $q$ is 1.
15: Load $f_3$ in register $g$.
16: Control-add $f_3$ into register $f$ if the third qubit of register $q$ is 1.
17: Load $f_4$ in register $g$.
18: Control-add $f_4$ into register $f$ if the fourth qubit of register $q$ is 1.

$\triangleright$ Define validity of the candidates.

19: Load the maximum weight (10 kg in the example) in register $g$.
20: Use a comparator with register $w$ to define the validity of a candidate in $v$.
21: Inverse the fitness $f$ of the invalid candidates (for which $v$ is 1).

$\triangleright$ Mark better candidates.

22: Load current maximum in register $g$.
23: Use a comparator with register $f$ to mark better candidates than the current one in register $r$.
24: Uncompute.

---

# References

1. Ahuja, A., Kapoor, S.: A quantum algorithm for finding the maximum, November 1999. arXiv:quant-ph/9911082
2. Boyer, M., Brassard, G., Hoeyer, P., Tapp, A.: Tight bounds on quantum searching (1996). https://doi.org/10.1002/(SICI)1521-3978(199806)46:4/5⟨493::AID-PROP493⟩3.0.CO;2-P

3. Chakrabarty, I., Khan, S., Singh, V.: Dynamic Grover search: applications in recommendation systems and optimization problems. Quant. Inf. Process. **16**(6), 153 (2017). https://doi.org/10.1007/s11128-017-1600-4
4. Cuccaro, S., Draper, T., Kutin, S.: A new quantum ripple-carry addition circuit, October 2004. arXiv:quant-ph/0410184
5. Durr, C., Hoyer, P.: A quantum algorithm for finding the minimum. arXiv preprint arXiv:quant-ph/9607014 (1996)
6. Grover, L.K.: A fast quantum mechanical algorithm for database search (1996)
7. Muñoz-Coreas, E., Thapliyal, H.: T-count optimized design of quantum integer multiplication. CoRR abs/1706.05113 (2017)
8. Thapliyal, H., Ranganathan, N.: Design of efficient reversible logic-based binary and BCD adder circuits. J. Emerg. Technol. Comput. Syst. **9**(3), 17:1–17:31 (2013). https://doi.org/10.1145/2491682
9. Udrescu, M., Prodan, L., Vladutiu, M.: Implementing quantum genetic algorithms: a solution based on Grover's algorithm. In: Conference on Computing Frontiers (2006)

# Applications of Quantum Annealing

# Flight Gate Assignment with a Quantum Annealer

Tobias Stollenwerk[1]([⊠]) [iD], Elisabeth Lobe[1] [iD], and Martin Jung[2] [iD]

[1] High Performance Computing, Simulation and Software Technology,
German Aerospace Center (DLR), Linder Höhe, 51147 Köln, Germany
[2] Airport Research, Institute of Air Transport and Airport Research,
German Aerospace Center (DLR), Linder Höhe, 51147 Köln, Germany
{tobias.stollenwerk,elisabeth.lobe,m.jung}@dlr.de

**Abstract.** Optimal flight gate assignment is a highly relevant optimization problem from airport management. Among others, an important goal is the minimization of the total transit time of the passengers. The corresponding objective function is quadratic in the binary decision variables encoding the flight-to-gate assignment. Hence, it is a quadratic assignment problem being hard to solve in general. In this work we investigate the solvability of this problem with a D-Wave quantum annealer. These machines are optimizers for quadratic unconstrained optimization problems (QUBO). Therefore the flight gate assignment problem seems to be well suited for these machines. We use real world data from a mid-sized German airport as well as simulation based data to extract typical instances small enough to be amenable to the D-Wave machine. In order to mitigate precision problems, we employ bin packing on the passenger numbers to reduce the precision requirements of the extracted instances. We find that, for the instances we investigated, the bin packing has little effect on the solution quality. Hence, we were able to solve small problem instances extracted from real data with the D-Wave 2000Q quantum annealer.

**Keywords:** Quadratic assignment problem · Quantum annealing · QUBO · Airport planning

## 1 Introduction

Modern airport management requires a more holistic approach to the whole travel chain, aiming at a better situational awareness of airport landside processes and an improved resource management. One main key to achieve this is proactive passenger management [3]. Unlike common reactive approaches, proactive passenger management utilizes an early knowledge about the passengers' status and the expected situation in the terminal along with resulting system loads and resource deployment, e.g. using simulation techniques like in [9]. An appropriate and modern management also considers dependencies of airside and landside operations as well as costs and performance. One part of the proactive

© Springer Nature Switzerland AG 2019
S. Feld and C. Linnhoff-Popien (Eds.): QTOP 2019, LNCS 11413, pp. 99–110, 2019.
https://doi.org/10.1007/978-3-030-14082-3_9

passenger management is a proper flight gate assignment. Especially large connecting hub airports have to deal with the still growing demand of traffic and rising numbers of passengers and baggage that needs to be transferred between flights. Although there are several objective functions in the field of airport planning, we focus on reducing the total transit time for passengers in an airport [10]. The main goals are the increase of passenger comfort and punctuality. This problem is related to the quadratic assignment problem, a fundamental problem in combinatorial optimization whose standard formulation is NP-hard [5].

In general, most planning problems belong to the class of discrete optimization problems which are hard to solve classically. Therefore it is worth studying new hardware architectures, like quantum computers, which may outperform classical devices. The first commercially available quantum annealer, developed by the Canadian company D-Wave Systems, is a heuristic solver using quantum annealing for optimizing quadratic functions over binary variables without further constraints (QUBOs). A wide range of combinatorial optimization problems can be brought into such a QUBO format by standard transformations [8]. But usually these transformations produce overhead, e.g. by increasing the number of variables, the required connections between them or the value of the coefficients. It is important to assess the impact of this overhead in order to estimate the performance of future devices [14]. In this work, we investigate the solvability of the optimal flight gate assignment problem with a D-Wave 2000Q quantum annealer.

The paper is structured as follows: In Sect. 2 we formally introduce the problem preparing the mapping to QUBO, which is done in Sect. 3. The extraction of smaller, but representative problem instances is covered in Sect. 4. In Sect. 5 we present our results for solving these smaller problem instances with a D-Wave 2000Q quantum annealer.

## 2   Formal Problem Definition

Flight gate assignment was already addressed in different versions [7,12]. The following mainly corresponds to the formulation from [10] as a quadratic binary program with linear constraints.

### 2.1   Input Parameters

The typical passenger flow in an airport can usually be divided into three parts: After the airplane has arrived at the gate, one part of the passengers passes the baggage claim and leaves the airport. Other passengers stay in the airport to take connecting flights. These transit passengers can take up a significant fraction of the total passenger amount. The third part are passengers which enter the airport through the security checkpoint and leave with a flight. The parameters of the problem are summarized in Table 1.

**Table 1.** Input data for a flight gate assignment instance

| | |
|---|---|
| $F$ | Set of flights ($i \in F$) |
| $G$ | Set of gates ($\alpha \in G$) |
| $n_i^{\text{dep}}$ | Number of passengers which depart from the airport on flight $i$ |
| $n_i^{\text{arr}}$ | Number of passengers from flight $i$ which arrive at the airport |
| $n_{ij}$ | Number of transfer passengers which arrive on flight $i$ and depart with flight $j$ |
| $t_\alpha^{\text{arr}}$ | Time it takes for a passenger to get from gate $\alpha$ to baggage claim |
| $t_\alpha^{\text{dep}}$ | Time it takes for a passenger to get from check-in to gate $\alpha$ |
| $t_{\alpha\beta}$ | Time it takes to get from gate $\alpha$ to gate $\beta$ |
| $t_i^{\text{in}}$ | Arrival time of flight $i$ |
| $t_i^{\text{out}}$ | Departure time of flight $i$ |
| $t^{\text{buf}}$ | Buffer time between two flights at the same gate |

## 2.2   Variables and Objective

Assignment problems can easily be represented with binary variables indicating whether or not a resource is assigned to a certain facility. The variables form a matrix indexed over the resources and the facilities. The binary decision variables are $x \in \{0,1\}^{F \times G}$ with

$$x_{i\alpha} = \begin{cases} 1, & \text{if flight } i \in F \text{ is assigned to gate } \alpha \in G, \\ 0, & \text{otherwise.} \end{cases}$$

Like already stated, the passenger flow divides into three parts and so does the objective function: The partial sums of the arriving, the departing and the transfer passengers sum up to the total transfer time of all passengers. For the arrival part we get a contribution of the corresponding time $t_\alpha^{\text{arr}}$ for each of the $n_i^{\text{arr}}$ passengers if flight $i$ is assigned to gate $\alpha$. Together with the departure part, which is obtained analogously, the linear terms of the objective are

$$T^{\text{arr/dep}}(x) = \sum_{i\alpha} n_i^{\text{arr/dep}} t_\alpha^{\text{arr/dep}} x_{i\alpha}.$$

The contribution of the transfer passengers is the reason for the hardness of the problem: Only if flight $i$ is assigned to gate $\alpha$ and flight $j$ to gate $\beta$ the corresponding time is added. This results in the quadratic term

$$T^{\text{trans}}(x) = \sum_{i\alpha j\beta} n_{ij} t_{\alpha\beta} \, x_{i\alpha} x_{j\beta}.$$

The total objective function is

$$T(x) = T^{\text{arr}}(x) + T^{\text{dep}}(x) + T^{\text{trans}}(x). \tag{1}$$

## 2.3   Constraints

Not all binary encodings for $x$ form valid solutions to the problem. There are several further restrictions which need to be added as constraints. In this model a flight corresponds to a single airplane arriving and departing at a single gate. It is obvious, that every flight can only be assigned to a single gate, therefore we have

$$\sum_{\alpha} x_{i\alpha} = 1 \qquad \forall i \in F. \tag{2}$$

Furthermore it is clear that no flight can be assigned to a gate which is already occupied by another flight at the same time. These forbidden flight pairs with overlapping time slots can be aggregated in

$$P = \left\{ (i,j) \in F^2 : t_i^{in} < t_j^{in} < t_i^{out} + t^{buf} \right\}.$$

The resulting linear inequalities $x_{i\alpha} + x_{j\alpha} \leq 1$ are equivalent to the quadratic constraints

$$x_{i\alpha} \cdot x_{j\alpha} = 0 \qquad \forall (i,j) \in P \ \forall \alpha \in G.$$

## 3   Mapping to QUBO

QUBOs are a special case of integer programs minimizing a quadratic objective function over binary variables $x \in \{0,1\}^n$. The standard format is the following

$$q(x) = x^\top Q x = \sum_{j=1}^{n} Q_{jj} x_j + \sum_{\substack{j,k=1 \\ j<k}}^{n} Q_{jk} x_j x_k \tag{3}$$

with an upper-triangular quadratic matrix $Q \in \mathbb{R}^{n \times n}$. While the presented objective function already follows this format the constraints need to be reduced which is shown in this section.

### 3.1   Penalty Terms

The standard way to reduce constraints, like already shown in e.g. [14], is to introduce terms penalizing those variable choices that violate the constraints. Just in these cases a certain positive value is added to the objective function to favor valid configurations while minimizing. The quadratic terms

$$C^{one}(x) = \sum_i \left( \sum_\alpha x_{i\alpha} - 1 \right)^2, \quad C^{not}(x) = \sum_\alpha \sum_{(i,j)\in P} x_{i\alpha} x_{j\alpha}$$

fulfill

$$C^{one/not} \begin{cases} > 0, & \text{if constraint is violated,} \\ = 0, & \text{if constraint is fulfilled,} \end{cases}$$

and therefore are suitable penalty terms which can be combined with the objective function. Since the benefit in the objective function in case of an invalid variable choice should not exceed the penalty, two parameters $\lambda_{one}, \lambda_{not} \in \mathbb{R}_+$ need to be introduced:

$$q(x) = T(x) + \lambda^{one} C^{one}(x) + \lambda^{not} C^{not}(x).$$

In theory these parameters could be set to infinity, but in practice this is not possible and they have to be chosen carefully.

## 3.2  Choosing Penalty Weights

The parameters $\lambda^{one}$ and $\lambda^{not}$ need to be large enough to ensure that a solution always fulfills the constraints. However due to precision restrictions of the D-Wave machine it is favorable to choose them as small as possible. In this section we present two different possibilities to obtain suitable values. Besides a worst case analysis for each single constraint a bisection algorithm which iteratively checks penalty weights against the solution validity is presented. In the following we call a variable to be activated if it is set to one.

**Worst Case Estimation.** The minimal contribution of $C^{one}$ and $C^{not}$ when breaking the corresponding constraint is just one. Therefore the corresponding minimal penalties are $\lambda^{one}$ and $\lambda^{not}$. Since the objective function and $C^{not}$ only contain non-negative coefficients, just $C^{one}$ enforces some of the variables to be set to one. Therefore activating more than one $x_{i\alpha}$, for one flight $i$, does not improve the objective value. Hence $\lambda^{one}$ just needs to exceed the benefit of deactivating a single variable in a valid solution. This variable usually appears in several summands of the objective function. Hence in the worst case the objective could be reduced by the sum of all coefficients of monomials including this variable, which is

$$T_{i\alpha}(x) := n_i^{dep} t_\alpha^{dep} + n_i^{arr} t_\alpha^{arr} + \sum_j n_{ij} \sum_\beta t_{\alpha\beta} x_{j\beta}.$$

Assuming the penalty is chosen large enough it will also be sufficient for all gates $\beta$ appearing in $T_{i\alpha}$. Therefore in the last part of the sum, for every flight $j$, just one gate $\beta$ needs to be taken into account for which $x_{j\beta}$ is one. In the worst case this is the one with the maximal time $t_{\alpha\beta}$, which results in

$$T^{one} := \max_{i,\alpha} \left( n_i^{dep} t_\alpha^{dep} + n_i^{arr} t_\alpha^{arr} + \max_\beta t_{\alpha\beta} \sum_j n_{ij} \right).$$

Considering $C^{not}$, it is not preferable to add an additional flight pair that is forbidden. But if the penalty is not large enough, assigning one flight $i$ to a gate $\gamma$ although this gate is already occupied by another flight $j$ rather than using the different gate $\alpha$ could reduce the objective value. This means $x_{i\gamma} = 1$ might be preferred to $x_{i\alpha} = 1$ although $(i,j) \in P$ and $x_{j\gamma} = 1$. The resulting benefit can be calculated from the difference of $T_{i\alpha}$ and $T_{i\gamma}$. For the estimation of the worst case this can be simplified by taking the maximum transfer time for $T_{i\alpha}$ and the minimum for $T_{i\gamma}$:

$$T^{\text{not}} := \max_{i,\alpha,\gamma} \left( \left( n_i^{\text{dep}} t_\alpha^{\text{dep}} + n_i^{\text{arr}} t_\alpha^{\text{arr}} + \max_\beta t_{\alpha\beta} \sum_j n_{ij} \right) \right.$$

$$\left. - \left( n_i^{\text{dep}} t_\gamma^{\text{dep}} + n_i^{\text{arr}} t_\gamma^{\text{arr}} + \min_\beta t_{\gamma\beta} \sum_j n_{ij} \right) \right)$$

$$= \max_{i,\alpha,\gamma} \left( \left( n_i^{\text{dep}} t_\alpha^{\text{dep}} - n_i^{\text{dep}} t_\gamma^{\text{dep}} \right) + \left( n_i^{\text{arr}} t_\alpha^{\text{arr}} - n_i^{\text{arr}} t_\gamma^{\text{arr}} \right) + \max_\beta \left( t_{\alpha\beta} - t_{\gamma\beta} \right) \sum_j n_{ij} \right).$$

All in all using $\lambda^{\text{one}} = T^{\text{one}} + \varepsilon$ and $\lambda^{\text{not}} = T^{\text{not}} + \varepsilon$ for some $\varepsilon > 0$ ensures that the minimum of $q$ satisfies the constraints. The provided boundaries can be calculated easily, but unfortunately they may take pretty large values depending on the given parameters. Since usually the worst case is also not the most probable it is a very rough estimation and there is some room for improvement.

**Bisection Method.** If the constraints are independent of each other, a simple bisection method can be used to find an approximation of the boundary between valid and invalid penalty values. In the course of this, all but one penalty is fixed, while the bisection is employed in one dimension at a time. These fixed penalty values as well as an upper starting point of the bisection method need to yield valid solutions. For given values of the penalties, a classical solver like SCIP is used to find the exact solution [6]. But it is imaginable that also the D-Wave machine itself could be used. For very small instances it might happen that one of the constraints becomes redundant and therefore is always fulfilled if the other one is. In these cases just the remaining constraint needs to be evaluated. This method, may lead to much smaller values than the worst case estimation. However it requires solving the problem several times exactly for each instance. Therefore this approach is not viable in an operational setting.

## 4    Instance Extraction and Generation

From [9], applying agent based simulation techniques of [1], we extract the data to estimate the transfer times $t^{\text{buf}}$, $t_\alpha^{\text{dep}}$, $t_\alpha^{\text{arr}}$ $\forall \alpha \in G$ and $t_{\alpha\beta}$ $\forall (\alpha, \beta) \in G^2$. The second part of our data source consists of a flight schedule of a mid-sized German airport for a full day. This gives us the number of passengers $n_i^{\text{dep}}$, $n_i^{\text{arr}}$ $\forall i \in F$ and $n_{ij}$ $\forall (i,j) \in F^2$. The resulting problem instance is depicted in Fig. 1.

### 4.1    Preprocessing

Since the given data set contains a slice of a flight schedule cut off at one day, there are flights which either do not have an arrival or a departure time. These

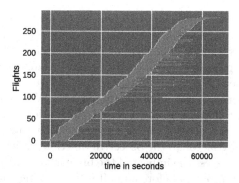

**Fig. 1.** Left: Time at the airport for all flights during a whole day on a mid-sized German airport. Right: Graph representing the transfer passengers for a whole day. Each vertex is a flight, each edge are transfer passengers between two flights.

flights are removed. Furthermore for some of the airplanes, the standing time is so long that it is reasonable to assume the airplane is moved away from the gate for some time before returning to a possibly different gate. Hence, we considered all flights with a time at the airport above 120 min as two separate flights. In order to extract typical, hard instances from the data, we removed all flights with no transfer passengers. The remaining instance now consists of 89 flights with 80 transfers and over 35 gates. However, since this corresponds to more than 3000 binary variables further subdivision is needed to make the problem amenable to the current D-Wave 2000Q hardware.

A way to achieve this becomes apparent when visualizing the dependencies among flights with transfer passengers in a graph as in the right hand side of Fig. 1. It can be observed that the graph is divided into several connected components, various smaller ones with 3 to 11 flights and a single larger one. The time intervals are much more distributed. Therefore extracting these special subgraphs provides suitable instances to be tested. In addition to using the connected components of the transfer passenger graph, we randomly cut the largest connected component to create larger test instances. This results in 163 instances with number of flights and gates from 3 to 16 and 2 to 16, respectively. The set of these instances will be denoted by $\mathcal{I}_{CC}$.

An alternative to using only flights with transfer passengers, is to use only flights within a certain time interval. However, this option has two major disadvantages: In such a time slot most of the flight intervals overlap mutually and therefore each flight requires a different gate. Also, in our data set, these flights do have almost no transferring passengers. Both issues simplify the subproblems severely. Therefore this alternative is not pursued in this work.

## 4.2 Bin Packing

As we will see, some of the instances from $\mathcal{I}_{CC}$ are small enough to fit on the D-Wave 2000Q machine. However, they exhibit a strong spread of the coefficients in the QUBO. It is known that this can suppress the success probability due to the limited precision of the D-Wave machine [15]. Therefore, we tried to reduce the spread of the coefficients while retaining the heart of the problem as best as possible. First, we mapped the passenger numbers to natural numbers in $\{0, 1, \ldots, N_p\}$ with bin packing, where $N_p$ is the number of bins. Moreover we mapped the transfer times to random natural numbers from $\{1, \ldots, N_t\}$. This is reasonable, since the original time data was drawn from a simulation data, which showed similar behavior. The mapping of the maximum transfer time in the instance to $N_t$ introduces a scaling factor to the objective function, which is irrelevant for the solution. In order to assess the impact of the bin packing for the number of passengers, we solved each problem before and after bin packing with the exact solver SCIP. Let the solutions before and after bin packing be denoted as $\mathbf{x}$ and $\hat{\mathbf{x}}$. The approximation ratio is then given by the $R = \frac{T(\hat{\mathbf{x}})}{T(\mathbf{x})}$. Where $T$ is the objective function (1) before bin packing. If $R = 1$, the bin packing has no effect on the solution quality. Figure 2 shows the approximation ratio for various bin packed instances. We used values of $N_p \in \{2, 3, 6, 7, 8, 9, 10\}$ and $N_t \in \{2, 3, 6, 10\}$ and all combinations thereof. Moreover, we restrict ourselves to instances with $|F| \cdot |G| < 100$. One can see, that the approximation ratio is close to one for the majority of the bin packed instances. Meaning, the bin packing has little effect on the solution quality, at least for the instances we investigated.

For the investigation of quantum annealing, we restrict ourselves to $N_p, N_t \in \{2, 3, 6, 10\}$. The corresponding instances will be denoted by $\mathcal{I}_{BP}$.

## 5   Quantum Annealing

### 5.1   Embedding

The hardware layout of the D-Wave 2000Q quantum annealer restricts the connections between binary variables to the so called Chimera graph [13]. In order to make problems with higher connectivity amenable to the machine, we employ minor-embedding [2]. This includes coupling various physical qubits together

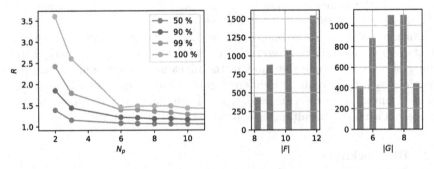

**Fig. 2.** Left: Approximation ratio for bin packed instances. The different curves show percentiles. Right: Distribution of number of flights and gates $|F|$, $|G|$.

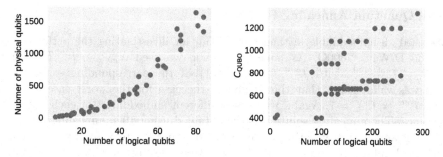

**Fig. 3.** Left: Number of logical and physical qubits on D-Wave 2000Q for instances $\mathcal{I}_{CC}$. Right: Maximum coefficient ratio of the QUBOs for instances $\mathcal{I}_{CC}$.

into one logical qubit, representing one binary variable, with a strong ferromagnetic coupling $J_F$ in order to ensure that all physical qubits have the same value after readout [16]. Since the constraint (2) introduces $|F|$ complete graphs of size $|G|$, we expect at most quadratic increase in the number of physical qubits with the number of logical qubits, which is $|F| \cdot |G|$. This is supported by our findings in the left hand side of Fig. 3. We calculated five different embeddings for each problem instance from $\mathcal{I}_{CC}$. With this, we were able to embed instances up to 84 binary variables.

## 5.2 Precision Requirements

The D-Wave 2000Q has a limited precision in specifying the linear and quadratic coefficients of the Ising model. As it was shown in [15], this can be an inhibiting factor for solving some problems on the D-Wave 2000Q. In order to assess the precision requirements for each instance, we introduce the maximum coefficient ratio for a QUBO like (3) as

$$C_{\text{QUBO}} = \frac{\max_{ij} |Q_{ij}|}{\min_{ij} |Q_{ij}|},$$

and for the corresponding embedded Ising model $\sum_i h_i s_i + \sum_{ij} J_{ij} s_i s_j$ as

$$C_{\text{Ising}} = \max \left\{ \frac{\max_i |h_i|}{\min_i |h_i|}, \frac{\max_{ij} |J_{ij}|}{\min_{ij} |J_{ij}|} \right\}.$$

The right hand side of Fig. 3 shows the maximum coefficient ratio for all QUBO from $\mathcal{I}_{CC}$. We used minimal sufficient penalty weights calculated by bisection as it was described in Sect. 3.2. The corresponding values for the embedded Ising model are orders of magnitude larger (not shown). Therefore the success probability for these instances is highly suppressed. However, this does not seem to be sufficient to reduce the precision requirements of the instances to an acceptable level. In order to mitigate the problem, we will use the bin packed instances $\mathcal{I}_{BP}$ for the remainder of this work.

## 5.3   Quantum Annealing Results

We used all embeddable instances from $\mathcal{I}_{BP}$ for investigating the performance
of the D-Wave 2000Q. As penalty weights, we used $\lambda_{\mathrm{one}} = f^{\mathrm{one}}T^{\mathrm{one}}$ and
$\lambda_{\mathrm{not}} = f^{\mathrm{not}}T^{\mathrm{not}}$, with $f^{\mathrm{one}}, f^{\mathrm{not}} \in \{\frac{1}{2}, 1\}$, if the corresponding exact solution was valid. Note, that this is always the case for the worst case estima-
tion $f^{\mathrm{one}} = f^{\mathrm{not}} = 1$. Again, we used 5 different embeddings for each problem
instance. The annealing solutions were obtained using 1000 annealing runs, no
gauging and majority voting as an un-embedding strategy for broken chains
of logical qubits. In order to calculate the time-to-solution with 99% certainty
we use $T_{99} = \frac{\ln(1-0.99)}{\ln(1-p)}T_{\mathrm{anneal}}$, where $T_{\mathrm{anneal}}$ is the annealing time, which we
fixed to $20\mu s$, and $p$ is the success probability. The latter is calculated by the
ratio of the number of runs where the optimal solution was found to the total
number of runs. The best annealing solution was compared to an exact solution
obtained with a MaxSAT solver [11]. Note, that this could be relaxed by accepting approximate solutions as long as they are valid. As expected, we found that
the time-to-solution increases with decreasing solution quality (not shown). As
intra-logical qubit coupling we used $J_F = -1$ in units of the largest coefficient of
the embedded Ising model of the problem instance at hand. The left hand side
of Fig. 4 shows the dependence of the success probability on the choice of $J_F$ for
a single instance from $\mathcal{I}_{BP}$. As expected, for large $J_F$ the success probability is
suppressed due to increased precision requirements and for very small $J_F$ the
logical qubit chains are broken (cf. [15]). The former is substantiated by the
right hand side of Fig. 4, where the success probability in dependence of $C_{\mathrm{Ising}}$
is shown for a fixed $J_F = -1$ in units of the largest coefficient of the embedded
Ising model. The success probability is suppressed for larger values of $C_{\mathrm{Ising}}$ due
to the increased precision requirements. The left hand side of Fig. 5 shows the
time-to-solution in dependence of the number of flights. There is an increase in
the time-to-solution with the number of flights, and therefore the problem size.
This can be explained by the increase in $C_{\mathrm{Ising}}$ with the number of flights as it
can be seen on the right hand side of Fig. 5.

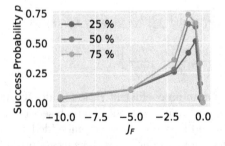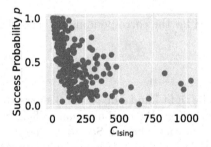

**Fig. 4.** Left: Success probability against the intra-logical qubit coupling in units of
the largest coefficient of the embedded Ising model. The data is for a representative
instance from $\mathcal{I}_{BP}$. The different colors represent the 25th, 50th and 75th percentiles.
Right: Maximum success probability against $C_{\mathrm{Ising}}$ for the instances from $\mathcal{I}_{BP}$. (Color
figure online)

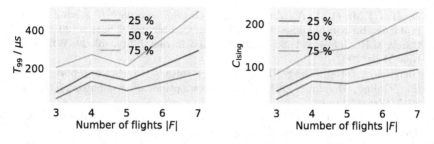

**Fig. 5.** Left: Success probability against the number of flights for the instances from $\mathcal{I}_{BP}$. Right: Maximum $C_{Ising}$ against the number of flights for the instances from $\mathcal{I}_{BP}$. The different colors represent the 25th, 50th and 75th percentiles. (Color figure online)

## 6    Conclusion

We showed, that the flight gate assignment problem can be solved with a quantum annealer with some restrictions. First, the size of the amenable problems is very small. Due to the high connectivity of the problem there is a large embedding overhead. Therefore a conclusive assessment of the scaling behavior is not possible at the moment. Future generations of quantum annealers with more qubits and higher connectivity are needed to investigate larger problems. Second, extracting problem instances directly from the data can lead to distributed coefficients in the resulting QUBOs. As a result, the success probability is mostly suppressed for these instances due to their high precision requirements. However, bin packing the coefficients can strongly decrease the precision requirements while retaining the heart of the problem.

For future work we leave the investigation of hybrid algorithms in order to recombine partial solutions and solve the whole problem (cf. [4]), the further investigation of the influence of bin packing on the solution quality, as well as the comparison of the performance to classical methods.

**Acknowledgments.** The authors would like to thank NASA Ames Quantum Artificial Intelligence Laboratory for their support during performing the experiments on the D-Wave 2000Q system, for many valuable discussions and the opportunity to use the D-Wave 2000Q machine at NASA Ames.

## References

1. Bonabeau, E.: Agent-based modeling: methods and techniques for simulating human systems. Natl. Acad. Sci. **99**, 7280–7287 (2002)
2. Cai, J., Macready, W.G., Roy, A.: A practical heuristic for finding graph minors. arXiv preprint, June 2014. http://arxiv.org/abs/1406.2741
3. Classen, A.B., Rudolph, F.: Proactive passenger management with a total airport management perspective. In: Transportation Research Forum - TRF 2015, pp. 1–19, März 2015. https://elib.dlr.de/96079/

4. Do, T.T.T.M., et al.: A hybrid quantum-classical approach to solving scheduling problems. In: Ninth Annual Symposium on Combinatorial Search (2016)
5. Garey, M.R., Johnson, D.S.: Computers and Intractability, vol. 29. WH Freeman, New York (2002)
6. Gleixner, A., et al.: The SCIP optimization suite 5.0. Technical report, Optimization Online, December 2017. http://www.optimization-online.org/DB_HTML/2017/12/6385.html
7. Haghani, A., Chen, M.C.: Optimizing gate assignments at airport terminals. Transp. Res. Part A: Policy Pract. **32**(6), 437–454 (1998)
8. Hammer, P.L., Rudeanu, S.: Boolean Methods in Operations Research and Related Areas, vol. 7. Springer, New York (2012). https://doi.org/10.1007/978-3-642-85823-9
9. Jung, M., Classen, A.B., Rudolph, F., Pick, A., Noyer, U.: Simulating a multi-airport region to foster individual door-to-door travel. In: Winter Simulation Conference, pp. 2518–2529. IEEE (2017). https://doi.org/10.1109/WSC.2017.8247980
10. Kim, S.H., Feron, E., Clarke, J.P., Marzuoli, A., Delahaye, D.: Airport gate scheduling for passengers, aircraft, and operations. J. Air Transp. **25**(4), 109–114 (2017). https://doi.org/10.2514/1.D0079
11. Kügel, A.: Improved exact solver for the weighted max-sat problem. Pos@ sat **8**, 15–27 (2010). https://www.uni-ulm.de/fileadmin/website_uni_ulm/iui.inst.190/Mitarbeiter/kuegel/maxsat.pdf
12. Mangoubi, R., Mathaisel, D.F.: Optimizing gate assignments at airport terminals. Transp. Sci. **19**(2), 173–188 (1985)
13. McGeoch, C.C., Wang, C.: Experimental evaluation of an adiabatic quantum system for combinatorial optimization. In: Proceedings of the ACM International Conference on Computing Frontiers, p. 23. ACM (2013)
14. Rieffel, E.G., Venturelli, D., O'Gorman, B., Do, M.B., Prystay, E.M., Smelyanskiy, V.N.: A case study in programming a quantum annealer for hard operational planning problems. Quant. Inf. Process. **14**(1), 1–36 (2015)
15. Stollenwerk, T., et al.: Quantum annealing applied to de-conflicting optimal trajectories for air traffic management. arXiv preprint (2017). https://arxiv.org/abs/1711.04889
16. Venturelli, D., Marchand, D.J.J., Rojo, G.: Quantum annealing implementation of job-shop scheduling. arXiv preprint, June 2015. http://arxiv.org/abs/1506.08479

# Solving Quantum Chemistry Problems with a D-Wave Quantum Annealer

Michael Streif[1]([⊠]), Florian Neukart[2,3], and Martin Leib[1]

[1] Data:Lab, Volkswagen Group, Ungererstr. 69, 80805 München, Germany
`michael.streif@volkswagen.de`
[2] LIACS, Leiden University, Niels Bohrweg 1, 2333 CA Leiden, Netherlands
[3] Volkswagen Group of America,
201 Post Street, San Francisco, CA 94108, USA

**Abstract.** Quantum annealing devices have been subject to various analyses in order to classify their usefulness for practical applications. While it has been successfully proven that such systems can in general be used for solving combinatorial optimization problems, they have not been used to solve chemistry applications. In this paper we apply a mapping, put forward by Xia et al. [25], from a quantum chemistry Hamiltonian to an Ising spin glass formulation and find the ground state energy with a quantum annealer. Additionally we investigate the scaling in terms of needed physical qubits on a quantum annealer with limited connectivity. To the best of our knowledge, this is the first experimental study of quantum chemistry problems on quantum annealing devices. We find that current quantum annealing technologies result in an exponential scaling for such inherently quantum problems and that new couplers are necessary to make quantum annealers attractive for quantum chemistry.

**Keywords:** Quantum computing · Quantum annealing · Quantum chemistry

## 1 Introduction

Since the seminal talk by Feynman [9] that jumpstarted the race for practical quantum computers, scientist have been dreaming of machines that can solve highly complicated quantum mechanical problems, which are inaccessible with classical resources. Molecules, such as caffeine, are already of such great complexity, that classical computers are incapable of simulating the full dynamics. Similar problems arise in the search for more efficient batteries, which is an important task for the upcoming electrification of traffic. Another prominent example is the simulation of photosynthesis processes in organic materials, which might lead to more efficient solar cells but is impossible to simulate due to the highly complex structure of the problem. The common ingredient which makes these problems so incredible hard to simulate is the exponentially growing Hilbert space. However a quantum computer, being a complicated quantum mechanical

© Springer Nature Switzerland AG 2019
S. Feld and C. Linnhoff-Popien (Eds.): QTOP 2019, LNCS 11413, pp. 111–122, 2019.
https://doi.org/10.1007/978-3-030-14082-3_10

system itself, harnesses the power of the exponentially growing Hilbert space. Therefore, quantum computing is generally believed to be able to find solutions to these complex and important problems. An important class of such quantum problems are electronic structure problems [15, 24]. There, the goal is to find the ground state energy of a quantum system consisting of electrons and stationary nuclei. To simulate these systems one has to solve the Schrödinger equation, $H |\Phi\rangle = E |\Phi\rangle$, where $H$ is the Hamiltonian describing the dynamics of the electrons in presence of stationary cores. Solving the Schrödinger equation can get unfeasible already when adding just another single electron to the system. Due to this exponentially increasing complexity of the problem solving this type of problem belongs to the most difficult calculations in both science and industry. State-of-the-art classical computers are able to solve such problems exactly for smaller molecules only. Various numerical methods such as Hartree-Fock (HF), quantum Monte Carlo (qMC), density functional theory (DFT) or configuration interaction methods (CI) were developed to give an approximate solution to these kinds of problems. In the future, quantum computers will be able to find exact solutions to these problems and thereby obtaining a deeper insight into nature.

In recent years early industrial incarnations of gate based quantum computers are appearing [6–8, 22] and small quantum chemistry algorithms have already been executed on these machines, such as the Variational Quantum Eigensolver (VQE) [21] or the Phase Estimation Algorithm (PEA) [13], which utilize the possibility to represent electrons in atomic orbitals as qubits on the quantum processor. Kandala et al. used transmon qubits to simulate molecular hydrogen, lithium hydride and beryllium oxide [12]. Hempel et al. used a trapped ion quantum computer to simulate molecular hydrogen and lithium hydride [10]. However, current gate model devices suffer from various shortcomings, e.g. a small number of qubits, errors caused by imperfect gates and qubits, and decoherence effects. These effects limit the coherence time and consequently the total number of gates which can be applied before decoherence effects destroy the potential for any quantum speedup. Research is therefore focusing on finding hardware-efficient, i.e. shallow circuits for solving electronic structure problems [3, 5, 12, 14].

Parallel to the development of gate based quantum computers there have been efforts to build quantum annealers on an industrial scale since the turn of the century [11] and the number of qubits in such devices have rapidly increased over the last few years. Quantum annealers use a quantum enhanced heuristic to find solutions to combinatorial optimization problems, such as traffic flow optimization [18] or cluster analysis [19]. In more recent time, quantum annealing systems were utilized to sample from a classical or quantum Boltzmann distribution, which enabled using such machines for machine learning purposes such as the training of deep neural networks [2] or reinforcement learning [16]. Despite their higher maturity with respect to gate based quantum computers, quantum annealers have, to the best of our knowledge not been used to solve quantum chemistry problems yet. In this present contribution, we follow an approach put forward by Xia et al. [25] to map electronic structure Hamiltonians to a classical spin glass and subsequently find the ground state with quantum annealing.

We start in Sect. 2 with a short introduction into quantum annealing followed by a general presentation of the electronic structure problem and a recapitulation of the approach proposed by [25] in Sect. 3. In Sect. 4, we present first results obtained with the D-Wave 2000Q machine for molecular hydrogen ($H_2$) and lithium hydride (LiH). In Sect. 5, we give technical details of the obtained results and, finally, in Sect. 6, we conclude with a summary of our findings and an outlook of possible future research directions.

## 2    Quantum Annealing in a Nutshell

Quantum annealing belongs to a class of meta-heuristic algorithms suitable for solving combinatorial optimization problems. The quantum processing unit (QPU) is designed to find the lowest energy state of a spin glass system, described by an Ising Hamiltonian,

$$H_{SG} = \sum_i h_i \sigma_z^i + \sum_{i,j} J_{ij} \sigma_z^i \sigma_z^j, \tag{1}$$

where $h_i$ is the on-site energy of qubit $i$, $J_{ij}$ are the interaction energies of two qubits $i$ and $j$, and $\sigma_z^i$ is the Pauli matrix acting on the $i$-th qubit. Finding the ground state of such a spin glass system, i.e. the state with lowest energy, is a NP problem. Thus, by mapping other NP problems onto spin glass systems, quantum annealing is able to find the solution of them. The idea of quantum annealing is to prepare the system in the ground state of a Hamiltonian which is known and easy to prepare, e.g. $H_X = \sum_i \sigma_x^i$. Then we change the Hamiltonian slowly such that it is the spin glass Hamiltonian at time T,

$$H(t) = \left(1 - \frac{t}{T}\right) H_X + \left(\frac{t}{T}\right) H_{SG}. \tag{2}$$

If $T$ is long enough, according to the adiabatic theorem, the system will be in the ground state of the spin glass Hamiltonian $H_{SG}$.

## 3    The Electronic Structure Problem and Its Mapping on an Ising Hamiltonian

### 3.1    The Electronic Structure Problem

The behaviour of electrons inside molecules is determined by their mutual interaction and the interaction with the positively charged nuclei. To describe the dynamics of the electrons or to find their optimal energetic configuration one has to solve the Schrödinger equation of this many-body quantum system. The

Hamiltonian for a system consisting of $M$ nuclei and $N$ electrons can be written in first quantization as

$$H = -\sum_i \frac{\nabla_i^2}{2} - \sum_A \frac{\nabla_A^2}{2M_A} - \sum_{i,A} \frac{Z_A}{|r_i - R_A|} + \sum_{i,j} \frac{1}{|r_i - r_j|} + \sum_{A,B}^{M} \frac{Z_A Z_B}{|R_A - R_B|},$$

(3)

where $r_i$ are the positions of the electrons, $M_A$, $R_A$ and $Z_A$ are the mass, position and charge in atomic units of the nuclei respectively. Using the second quantization, i.e. writing Eq. 3 in terms of fermionic creation and annihilation operators $a_i^\dagger$ and $a_j$ of a specific fermionic mode $i$ and $j$ with $\{a_i, a_j^\dagger\} = \delta_{ij}$ and applying the Born-Oppenheimer approximation, which assumes that the nuclei do not move due to their much greater mass ($M_A \gg 1$), we get the following Hamiltonian

$$H = \sum_{i,j} h_{ij} a_j^\dagger a_i + \frac{1}{2} \sum_{i,j,k,l} h_{ijkl} a_i^\dagger a_j^\dagger a_k a_l.$$

(4)

The parameters $h_{ij}$ and $h_{ijkl}$ are the one- and two-particle integrals for a specific, problem-dependent basis set $|\psi_i\rangle$, which has to be appropriately chosen,

$$h_{ij} = \langle \psi_i | \left( -\frac{\nabla_i^2}{2} - \sum_A \frac{Z_A}{|r_i - R_A|} \right) |\psi_j\rangle,$$

$$h_{ijkl} = \langle \psi_i \psi_j | \frac{1}{|r_i - r_j|} |\psi_k \psi_l\rangle.$$

(5)

Quantum devices utilize qubits, i.e. we have to find a qubit representation of the fermionic Hamiltonian. The Jordan-Wigner or Bravyi-Kitaev transformation [23] are the most prominent examples achieving this task and map the fermionic creation and annihilation operators to Pauli matrices, both leading to a qubit Hamiltonian of the form

$$H = \sum_{i,\alpha} h_\alpha^i \sigma_\alpha^i + \sum_{i,j,\alpha,\beta} h_{\alpha\beta}^{ij} \sigma_\alpha^i \sigma_\beta^j + \sum_{i,j,k,\alpha,\beta,\gamma} h_{\alpha\beta\gamma}^{ijk} \sigma_\alpha^i \sigma_\beta^j \sigma_\gamma^k + \dots$$

(6)

where $\sigma_{\alpha=x,y,z}^i$ are the Pauli matrices acting on a single qubit at site $i$. This qubit Hamiltonian now encodes the electronic structure problem we would like to solve by using a quantum annealer.

## 3.2   Formulation as an Ising spin glass

To find the ground state energy of an electronic structure problem with a quantum annealing device, it is necessary to find a representation of the problem, cf. Eq. 6, in the form of a classical spin glass system described by the Ising Hamiltonian, cf. Eq. 1. To accomplish this, we use the method proposed by Xia et al. [25]. In this section, we shortly recapitulate the method, for a detailed overview, cf. [25].

Terms containing $\sigma_x$ and $\sigma_y$ in the electronic structure Hamiltonian prohibit a direct embedding on a quantum annealer. To overcome this problem, the idea is to introduce $r$ ancillary qubits for all $n$ qubits of the original Hamiltonian respectively. Each Pauli operator is mapped to a diagonal operator in the classical subspace of the larger $(r \times n)$-qubit Hilbert space using the following relations:

$$\sigma_x^i \rightarrow \frac{1 - \sigma_z^{ij}\sigma_z^{ik}}{2}S(j)S(k) \qquad\qquad \sigma_y^i \rightarrow i\frac{\sigma_z^{ik} - \sigma_z^{ij}}{2}S(j)S(k)$$

$$\sigma_z^i \rightarrow \frac{\sigma_z^{ij} + \sigma_z^{ik}}{2}S(j)S(k) \qquad\qquad \mathbb{1} \rightarrow \frac{1 + \sigma_z^{ij}\sigma_z^{ik}}{2}S(j)S(k) \qquad (7)$$

In these mappings, $\sigma_{\alpha=x,y,z}^{ij}$ denote the Pauli matrices acting on the j-th ancillary qubit of the i-th original qubit. The mapped operators then reproduce the action of the former operators on the wavefunction in the original Hilbert space. The function $S(j)$ takes care of the sign of the state in the original Hilbert space. By increasing $r$, we are able to grasp more quantum effects and therefore are able to get a more precise estimation of the true ground state energy. In the following, we use this mapping and the proposed algorithm in [25] to find a classical spin glass approximation of our electronic structure Hamiltonian, which we then solve by using the methods of quantum annealing.

# 4    Results from the D-Wave 2000Q

In this section, we present first examples of electronic structure calculations done on a physical quantum annealing device, namely the D-Wave 2000Q machine. This quantum annealing device has 2048 qubits and 6016 couplers, arranged in a Chimera-type structure. Further information about the precise architecture and technical details can be found in [1].

We calculate the ground state energies of molecular hydrogen ($H_2$) and lithium hydride (LiH) for various interatomic distances. Moreover, we calculate the number of required qubits when using a quantum annealing device with limited connectivity. For an overview of our used methods, we refer to Sect. 5.

## 4.1    Molecular Hydrogen - $H_2$

We start with molecular hydrogen, $H_2$, which has already been the testbed for first electronic structure calculations on gate model devices [10,12]. With the two electrons of molecular hydrogen we have to account for 4 atomic orbitals. After calculating the fermionic Hamiltonian, a Bravyi-Kitaev transformation yields the qubit Hamiltonian,

$$\begin{aligned}
H_{H_2} =\ & g_1\mathbb{1} + g_2\sigma_z^0\sigma_z^2\sigma_z^3 + g_3\sigma_z^0\sigma_z^2 + g_4\sigma_z^2 + g_5\sigma_y^0\sigma_z^1\sigma_y^2\sigma_z^3 + g_6\sigma_z^0\sigma_z^1\sigma_z^2 + g_7\sigma_z^0\sigma_z^1 \\
& + g_8\sigma_z^1\sigma_z^3 + g_8\sigma_z^1\sigma_z^2\sigma_z^3 + g_{10}\sigma_z^0\sigma_z^1\sigma_z^2\sigma_z^3 + g_{11}\sigma_z^1 + g_{12}\sigma_z^0 + g_{13}\sigma_x^0\sigma_z^1\sigma_x^2 \\
& + g_{14}\sigma_x^0\sigma_z^1\sigma_x^2\sigma_z^3 + g_{15}\sigma_y^0\sigma_z^1\sigma_y^2.
\end{aligned} \qquad (8)$$

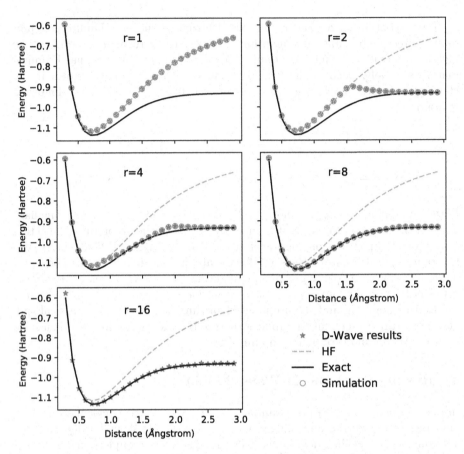

**Fig. 1.** Ground state energies of molecular hydrogen $H_2$ for various interatomic distances and different values of the scaling factor $r$. The red asterisks show the results from D-Wave 2000Q, the blue line shows the Hartree-Fock energy of the Hamiltonian, the black line the ground state energy obtained by the exact diagonalization of the molecular Hamiltonian, and the green circles show the simulated results of the classical $(r \times n)$-qubit Hamiltonian. As expected, by increasing the value of $r$, we increase the accuracy of the results. The D-Wave quantum annealing device closely reproduces the simulated results, meaning that it was able to find the right ground state energy for the given problem. For $r = 16$ we are very close to the exact results, where in this case we were not able to do the numerical calculation, thus we show experimental results only. In Fig. 2 we show the required qubits on the quantum processor for each of these plots. For all the experiments presented in this plot, we used an annealing time of $\tau = 100\,\mu s$ and 1000 annealing runs. (Color figure online)

In this expression, $g_i$ are parameters calculated from the one- and two-particle integrals, i.e. Eq. 5. These parameters depend on the interatomic distance $R$ between both hydrogen atoms, $g_i = g_i(R)$. As shown in [20], the first and third qubit in this Hamiltonian does not affect the population numbers of the Hartree-

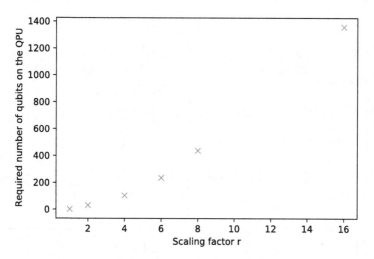

**Fig. 2.** Here we show the number of required qubits after embedding the classical $(r \times n)$-qubit Hamiltonian of molecular hydrogen on the real device for increasing scaling factors $r$.

Fock state $|\Psi_{\mathrm{HF}}\rangle$. Therefore both qubits are not important for finding the ground state energy and can be neglected in the remainder of this calculation, yielding the 2-qubit Hamiltonian

$$H_{\mathrm{H_2}} = g_0 \mathbb{1} + g_1\sigma_z^0 + g_2\sigma_z^1 + g_3\sigma_z^0\sigma_z^1 + g_4\sigma_x^0\sigma_x^1 + g_4\sigma_y^0\sigma_y^1 \tag{9}$$

Exact values for $g_i$ can be found in [20]. To find the ground state of this Hamiltonian, we map it on a $(r \times n)$-qubit Hamiltonian and find a classical 4-local spin glass representation in this larger Hilbert space. We then reduce the 4-local to 2-local terms by again introducing ancillary qubits and subsequently find the lowest eigenvalue of this 2-local classical spin glass Hamiltonian by embedding it on the Chimera graph of the D-Wave 2000Q machine. We repeat these experiments for different values of the interatomic distance and for different scaling factors $r$.

The results of these experiments are shown in Fig. 1. For $r = 1$, we merely remove any term in the Hamiltonian that contains $\sigma_x$ or $\sigma_y$ Pauli operators. As we start with a Hamiltonian which converged after a classical Hartree-Fock calculations, see Sect. 5, this case should provides us with Hartree-Fock energies. For $r = 2$, we have to account for 20 terms in total and 31 qubits on the QPU after embedding. For the largest scaling factor we used, $r = 16$, we have to use 1359 qubits on the QPU to account for 1490 terms in the 2-local spin glass version of the Hamiltonian.

In Fig. 2, we show the number of used qubits on the quantum annealing processor for different $r$. These numbers are the result of the mapping to a Hamiltonian containing only $\sigma_z$ operators, the mapping to two-local terms and the embedding on the chimera structure of the D-Wave 2000Q device. The required

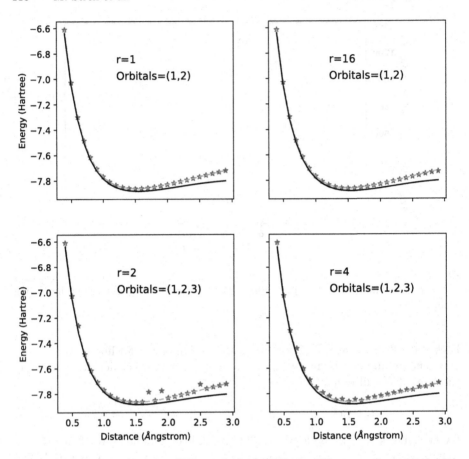

**Fig. 3.** Ground state energies of lithium hydride, LiH, for different active orbitals, different values of $r$ and various interatomic distances. LiH has 6 orbitals, here numbered from 0 to 5, plus spin degrees of freedom. The red asterisks show the results from D-Wave 2000Q, the blue line shows the Hartree-Fock energy of the Hamiltonian and the black line the exact ground state energy. For the sake of brevity, we do not show numerical results of the transformed Hamiltonian here. For the experiments with 2 orbitals, we used an annealing time of $\tau = 100\,\mu s$ and 1000 annealing runs, whereas for 3 orbitals, we used an annealing time of $\tau = 100\,\mu s$ and 9000 annealing runs. (Color figure online)

number of qubits approximately scales quadratically with the scaling factor. We note that this does not imply that the whole method scales quadratically when going to larger systems, as we might need an exponential increasing scaling factor for reaching an sufficient accuracy.

## 4.2   Lithium Hydride - LiH

To gauge the potential of the presented technique we start considering more complicated molecules such as lithium hydride. Lithium hydride has four electrons and we have to account for 12 orbitals in the minimal basis. We follow the same steps as for molecular hydrogen, i.e. we derive the qubit Hamiltonian of the problem using a Bravyi-Kitaev transformation of the fermionic Hamiltonian. We again are able to truncate the Hamiltonian to a smaller space by exploiting the fact that two qubits do not change its population numbers when starting in a Hartree-Fock state, leaving us with a 10-qubit Hamiltonian. Using the transformation given in Sect. 3.2, we map the Hamiltonian to a classical spin glass. In contrast to molecular hydrogen, for lithium hydride, due to limited number of qubits of the quantum hardware and the embedding overhead due to the limited connectivity, we are not able to take all atomic orbitals into account. Therefore we use an active space representation, i.e. we freeze orbitals and optimize the electron configuration in the remaining orbitals. After finding a 2-local representation of the remaining Hamiltonian and an embedding on the Chimera graph, we again find the ground state by the use of quantum annealing. For the sake of brevity, we do not state the full Hamiltonian here. In the following we present results for both various atomic orbitals and scaling factors.

These results can be found in Fig. 3. The results of all shown plots are very close to the initial Hartree-Fock energies, meaning that we were not able to improve from the initial Hartree-Fock energies, which was the starting point of our calculation. Additionally, as the problem gets more complicated, D-Wave was in some cases not able to find the true ground state energy of the transformed Hamiltonian. For a scaling factor $r = 4$ and 3 orbitals, we have to use 1558 qubits on the QPU. In our experiments, it was not possible to use more orbitals while $r \neq 1$.

## 5   Methods

In this section, we shortly summarize the technical details of our experiments. For both molecules, $H_2$ and LiH, we start by determining the fermionic Hamiltonian by calculating the one- and two-particle integrals, cf. Eq. 5 using the Psi4 module of Google's OpenFermion library [17]. We used the molecular wavefunctions from converged Hartree-Fock calculations obtained by using a minimal basis set, namely STO-3G, which are created from 3 Gaussian orbitals fitted to a single Slater-type orbital. We then apply a Bravyi-Kitaev transformation to map the second-quantized fermionic operators onto Pauli matrices to obtain a qubit representation of the problem. By using the method described in Sect. 3.2, we map the n-qubit Hamiltonian to a $(r \times n)$-qubit Hamiltonian. As D-Wave's implementation consists of 2-local terms only, we introduce ancillary qubits to find a 2-local representation of the Hamiltonian to embed the problem subsequently onto the Chimera graph structure of the quantum annealing device. To make sure that the embedding is optimal, we use the heuristic algorithm provided by D-Wave and generate 100 random embeddings for each bond length and find the ground state energy for each of these 100 embeddings with the D-Wave 2000Q. We then only keep the best solution, i.e. the solution with lowest energy. To compare our results with classical methods, we use the converged Hartree-Fock energies and the exact results, which we obtained by a numerical diagonalization of the qubit Hamiltonian.

# 6    Conclusion and Outlook

In this paper, we did a first examination of quantum chemistry problems on the D-Wave 2000Q by using an approach proposed by Xia et al. [25]. We were able to calculate the ground state energies of molecular hydrogen and lithium hydride with the current generation of the QPU. For molecular hydrogen, $H_2$, our ground state energy estimations were very close to the exact energies when going to scaling factor of $r = 16$. For achieving this accuracy, we already had to use a large fraction of available qubits on the quantum processor. We moreover showed a first scaling of this methods under real conditions and overheads of quantum annealing hardware. For lithium hydride, LiH, we were not able to reproduce closely the ground state energy with the currently available hardware. When accounting for 3 orbitals and using a scaling factor of $r = 4$, we already had to use 1558 qubits, which is a large fraction of available qubits. To summarize: the investigated method in general works, but it might be difficult to apply it to larger systems.

However, we give some further research ideas how quantum annealing devices could be applied to quantum chemistry problems in the nearer future. Quantum annealing devices which utilize interactions beyond the standard Ising Hamiltonian, i.e. beyond $\sigma_z \otimes \sigma_z$ interactions, could be helpful as they would allow to use a larger fraction of the Hilbert space. When having access to the right interactions, an efficient embedding onto quantum annealing processors could be feasible [4]. Another possibility is to use the recently announced new features of the D-Wave machine, such as the possibility to do reverse annealing or to stop the annealing process in an intermediate point of the adiabatic evolution, i.e. between start and final Hamiltonian. This may be utilized for getting access to terms which are non-diagonal in the computational basis and in the end could enable to sample from low-lying energy states. Together with a classical subroutine, one could find the solution of the problem. Another possibility could be to use the D-Wave machine to calculate Hartree-Fock, i.e. approximate energies of the problem. Together with a classical loop, quantum annealing devices could be used to estimate the Hartree-Fock energy of large molecules, which would be unfeasible with classical resources. Another promising alternative is to use machine learning to improve the found ground state energies.

**Acknowledgments.** We thank VW Group CIO Martin Hofmann and VW Group Region Americas CIO Abdallah Shanti, who enable our research. Any opinions, findings, and conclusions expressed in this paper do not necessarily reflect the views of the Volkswagen Group.

# References

1. Technical description of the D-wave quantum processing unit, 09–1109a-e
2. Adachi, S.H., Henderson, M.P.: Application of quantum annealing to training of deep neural networks. arXiv preprint arXiv:1510.06356 (2015)
3. Babbush, R., Berry, D.W., McClean, J.R., Neven, H.: Quantum simulation of chemistry with sublinear scaling to the continuum. arXiv preprint arXiv:1807.09802 (2018)
4. Babbush, R., Love, P.J., Aspuru-Guzik, A.: Adiabatic quantum simulation of quantum chemistry. Sci. Rep. **4**, 6603 (2014)
5. Babbush, R., Wiebe, N., McClean, J., McClain, J., Neven, H., Chan, G.K.: Low depth quantum simulation of electronic structure. arXiv preprint arXiv:1706.00023 (2017)
6. Barends, R., et al.: Digitized adiabatic quantum computing with a superconducting circuit. Nature **534**(7606), 222 (2016)
7. Debnath, S., Linke, N.M., Figgatt, C., Landsman, K.A., Wright, K., Monroe, C.: Demonstration of a small programmable quantum computer with atomic qubits. Nature **536**(7614), 63 (2016)
8. DiCarlo, L., et al.: Demonstration of two-qubit algorithms with a superconducting quantum processor. Nature **460**(7252), 240 (2009)
9. Feynman, R.P.: Simulating physics with computers. Int. J. Theor. Phys. **21**(6–7), 467–488 (1982)
10. Hempel, C., et al.: Quantum chemistry calculations on a trapped-ion quantum simulator. arXiv preprint arXiv:1803.10238 (2018)
11. Johnson, M.W., et al.: Quantum annealing with manufactured spins. Nature **473**(7346), 194 (2011)
12. Kandala, A., et al.: Hardware-efficient variational quantum eigensolver for small molecules and quantum magnets. Nature **549**(7671), 242 (2017)
13. Kitaev, A.Y.: Quantum measurements and the Abelian stabilizer problem. arXiv preprint quant-ph/9511026 (1995)
14. Kivlichan, I.D., et al.: Quantum simulation of electronic structure with linear depth and connectivity. Phys. Rev. Lett. **120**(11), 110501 (2018)
15. Lanyon, B.P., et al.: Towards quantum chemistry on a quantum computer. Nat. Chem. **2**(2), 106 (2010)
16. Levit, A., Crawford, D., Ghadermarzy, N., Oberoi, J.S., Zahedinejad, E., Ronagh, P.: Free energy-based reinforcement learning using a quantum processor. arXiv preprint arXiv:1706.00074 (2017)
17. McClean, J.R., et al.: OpenFermion: the electronic structure package for quantum computers. arXiv preprint arXiv:1710.07629 (2017)
18. Neukart, F., Von Dollen, D., Compostella, G., Seidel, C., Yarkoni, S., Parney, B.: Traffic flow optimization using a quantum annealer. Front. ICT **4**, 29 (2017)
19. Neukart, F., Von Dollen, D., Seidel, C.: Quantum-assisted cluster analysis on a quantum annealing device. Front. Phys. **6**, 55 (2018)
20. O'Malley, P., et al.: Scalable quantum simulation of molecular energies. Phys. Rev. X **6**(3), 031007 (2016)
21. Peruzzo, A., et al.: A variational eigenvalue solver on a photonic quantum processor. Nature Commun. **5**, 4213 (2014)
22. Reagor, M., et al.: Demonstration of universal parametric entangling gates on a multi-qubit lattice. Sci. Adv. **4**(2), eaao3603 (2018)

23. Seeley, J.T., Richard, M.J., Love, P.J.: The Bravyi-Kitaev transformation for quantum computation of electronic structure. J. Chem. Phys. **137**(22), 224109 (2012)
24. Whitfield, J.D., Biamonte, J., Aspuru-Guzik, A.: Simulation of electronic structure Hamiltonians using quantum computers. Mol. Phys. **109**(5), 735–750 (2011)
25. Xia, R., Bian, T., Kais, S.: Electronic structure calculations and the Ising Hamiltonian. J. Phys. Chem. B **122**, 3384–3395 (2017)

# Solving Large Maximum Clique Problems on a Quantum Annealer

Elijah Pelofske[1]([✉]), Georg Hahn[2], and Hristo Djidjev[1]

[1] Los Alamos National Laboratory, Los Alamos, NM 87545, USA
{epelofske,djidjev}@lanl.gov
[2] Lancaster University, Lancaster LA1 4YW, UK
ghahn@cantab.net

**Abstract.** Commercial quantum annealers from D-Wave Systems can find high quality solutions of quadratic unconstrained binary optimization problems that can be embedded onto its hardware. However, even though such devices currently offer up to 2048 qubits, due to limitations on the connectivity of those qubits, the size of problems that can typically be solved is rather small (around 65 variables). This limitation poses a problem for using D-Wave machines to solve application-relevant problems, which can have thousands of variables. For the important Maximum Clique problem, this article investigates methods for decomposing larger problem instances into smaller ones, which can subsequently be solved on D-Wave. During the decomposition, we aim to prune as many generated subproblems that don't contribute to the solution as possible, in order to reduce the computational complexity. The reduction methods presented in this article include upper and lower bound heuristics in conjunction with graph decomposition, vertex and edge extraction, and persistency analysis.

**Keywords:** Branch-and-bound · Decomposition · D-Wave ·
Graph algorithms · Maximum clique · Optimization ·
Quantum annealing

## 1 Introduction

Quantum annealers such as the commercially available ones manufactured by D-Wave [9] aim to solve quadratic unconstrained binary optimization (QUBO) problems by looking for a minimum-energy state of a quantum system implemented on the hardware. Notably, due to the stochastic nature of annealing, the approximations of the global minimum returned by D-Wave are not always identical, while the solving time is independent of the problem instance. The class of QUBO problems includes many well-known NP-hard problems [19], for which no classical polynomial algorithms are known. In this article, we focus on the NP-hard problem of finding a maximum clique in a graph.

© Springer Nature Switzerland AG 2019
S. Feld and C. Linnhoff-Popien (Eds.): QTOP 2019, LNCS 11413, pp. 123–135, 2019.
https://doi.org/10.1007/978-3-030-14082-3_11

The D-Wave quantum processing unit is designed to minimize the *Hamiltonian*

$$H = \sum_{i \in V} a_i x_j + \sum_{(i,j) \in E} a_{ij} x_i x_j, \tag{1}$$

where $V = \{1, \ldots, n\}$, $\{x_i \mid i \in V\}$ is a set of $n$ binary variables, and $E \subseteq V \times V$ describes quadratic interactions between them. The coefficients $a_i$ are the weights of the linear terms, and $a_{ij}$ are the weights of the quadratic terms (*couplers*). If in (1) we have $x_i \in \{0, 1\}$ for all $i \in V$, the Hamiltonian is called a *QUBO*, whereas if $x_i \in \{-1, +1\}$ for all $i \in V$, it is called an *Ising* Hamiltonian. The Hamiltonian for the maximum clique problem we consider is a QUBO.

The *maximum clique problem (MaxClique)* is defined as the problem of finding a largest *clique* (a completely connected subgraph) in a graph $G = (V, E)$. The *clique number* is the size of the maximum clique. In [6], the QUBO formulation of the maximum clique problem is given as

$$H = -A \sum_{i \in V} x_i + B \sum_{(i,j) \in \overline{E}} x_i x_j, \tag{2}$$

provided $B/A$ is sufficiently large, where the choice $A = 1$ and $B = 2$ ensures equivalence of the maximum clique problem and the Hamiltonian in (2), and $\overline{E}$ is the edge set of the complement graph of $G$.

The D-Wave annealer has a fixed number of operational qubits which are arranged in a certain graph structure, consisting of a lattice of bipartite cells, on a physical chip [6]. D-Wave 2X has 1152 physical qubits, while the newest D-Wave 2000Q model has up to 2048 physical qubits. For many NP-hard problems, however, the connectivity structure of a QUBO (that is, the graph defined by its non-zero coefficients $a_{ij}$) does not match the specific connectivity of the qubits on the D-Wave chip. In particular, this is the case for general instances of MaxClique. This problem can be alleviated by computing a *minor embedding* of a complete graph onto the D-Wave architecture, thus guaranteeing that any QUBO connectivity can be mapped onto the D-Wave chip. In such an embedding, a connected set of physical qubits is identified to act as one logical qubit, which further reduces the number of available qubits [8]. Even if the connectivity structure required to solve a QUBO matches the one of D-Wave, the scaling of problem instances quickly exhausts the capacity of the D-Wave chip. In both cases, classical preprocessing can attempt to split up large problem instances into smaller ones, which can be solved on D-Wave. General purpose approaches for QUBO decomposition exist [3], even though they are not guaranteed to always work for arbitrary QUBO instances.

Special purpose techniques exist for decomposing certain NP-hard problems. This paper investigates a decomposition algorithm for finding a maximum clique in a graph [6], where the problem reduction does not result in a loss of information regarding the maximum clique of the original problem. The decomposition algorithm works by recursively splitting a graph at each level in such a way that the maximal clique is guaranteed to be contained in either of the two generated subgraphs. To reduce the computational complexity, we attempt in this article to identify and exclude those generated subgraphs that cannot contain a maximum clique. This is achieved by computing bounds on the maximal clique size at each iteration. The purpose of this article is to investigate several such techniques with the aim of demonstrating the viability of solving larger optimization problems with D-Wave than previously possible.

Decomposition has already been suggested to solve several NP-hard problems in [28], and since used for NP-hard problems such as domination-type problems and graph coloring in [25]. For the Maximum Independent Set problem, an equivalent formulation of MaxClique, polynomial-time algorithms are known for many special graph classes [10,15,21], including several algorithms relying on graph decomposition [7,14].

The article is structured as follows. Section 2 considers the problem of decomposing a MaxClique instance into smaller subproblems which fit D-Wave. The efficiency of the decomposition algorithm relies on the computation of upper and lower bounds on the clique size of the subgraph it generates, and a variety of techniques for computing such bounds is presented in Sect. 3, including subgraph extractions methods. Section 4 contains an experimental analysis of the sensitivity of our decomposition approach on the chosen subgraph extraction techniques and on the bounds of the clique size, as well as an analysis of the effectiveness of these techniques for very sparse and very dense graphs compared to earlier work. The article concludes with a discussion in Sect. 5.

## 2   Maximum Clique Decomposition

We use the CH-partitioning introduced in [11], see also [6], in order to split a large input graph $G = (V, E)$ into smaller subgraphs on which a maximum clique is found. We next briefly review the method.

### 2.1   CH-partitioning

The CH-partitioning algorithm works as follows. We iteratively select a single vertex $v \in V$ and extract the subgraph $G_v$ containing all neighbors of $v$ and all edges between them. Afterwards, $v$ and all edges adjacent to $v$ are removed from $G$, thus creating another new graph denoted as $G'$. This is visualized in Fig. 1. The clique number $\omega(G)$ of $G$ is equal to $\min(\omega(G_v) + 1, \omega(G'))$.

Before decomposing $G'$ and $G_v$ further in a similar fashion, our algorithm computes lower and upper bounds on the maximum clique sizes contained in $G'$ and $G_v$. This is helpful for two reasons. First, if the current best lower bound is less (worse) than the lower bound of a newly generated subgraph, we can update the current best lower bound accordingly. Second, if the upper bound for $G'$ or $G_v$ is less than the best current lower bound for the maximum clique, we can prune that subgraph and the branch of the decomposition tree associated with it. Our aim is to prune as many generated subproblems as possible, and Sect. 3 highlights the approaches we explore.

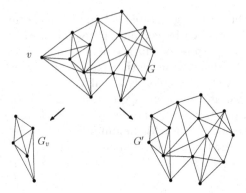

**Fig. 1.** Illustration of the vertex splitting at a single vertex $v$.

The algorithm stops splitting any generated subgraph as soon as it can be embedded on the D-Wave hardware, meaning if the subgraph size is at most 46 (65) vertices in case of D-Wave 2X (D-Wave 2000Q). The solution time of problems that fit the D-Wave hardware is essentially constant (anneals only without post-processing), so we are primarily concerned with the number of generated subproblems and the effectiveness of our pruning approaches. However, we do take into account that the success rate of annealing may depend on the graph structure and we compute different times-to-solutions for different graph types.

The proposed algorithm can be characterized as a branch-and-bound (B&B) algorithm, a class of algorithms often used to compute a solution of NP-hard optimization problems. As such, B&B has been previously used in exact classical algorithms for solving MaxClique, e.g., [5, 22]. However, these algorithms explore the search space in a quite different way than the one studied in this paper, as they work bottom up, starting with a single-vertex clique and expanding it as much as possible. In comparison, our algorithm is top-down, starting with the original graph and splitting it down to smaller graphs, which are examined for a maximum clique. Such structure allows for the generation of a set of smaller size MaxClique problems which fit in and are solvable on the quantum processing unit directly.

## 2.2  Vertex Choice

The algorithm of Sect. 2.1 offers one more tuning parameter: the procedure for selecting the vertex $v$ that is used in each iteration to split the current graph instance $G$ into two new graphs. Possible choices include:

  (i) a vertex $v$ of lowest degree.
 (ii) a vertex $v$ of median degree.
(iii) a vertex $v$ chosen at random.
(iv) a vertex $v$ of highest degree.

(v) a vertex $v$ in the $k$-core of lowest degree. The $k$-core of a graph $G = (V, E)$ is a maximal subgraph of $G$ in which every vertex has a degree of at least $k$. Specifically, we iteratively extract the $k$-core of sizes $k \in \{1, \ldots, |G|\}$ and stop at the first $k$ leading to a $k$-core which is not the original graph. We then choose $v$ as a vertex which was removed in that $k$-core reduction step.

(vi) a lowest degree vertex $v$ whose extracted subgraph $G_v$, defined in Sect. 2.1, is of lowest density.

In any of the above cases, if multiple vertices satisfy the selection criterion, the vertex $v$ which is extracted is chosen at random. The above vertex selection approaches are experimentally explored in Sect. 4.1.

# 3  Pruning of Subproblems

Next we describe two pruning approaches we use in order to reduce the number of subgraphs generated in the algorithm of Sect. 2. Section 3.1 describes bounds which can be used to prune entire subgraphs. Section 3.2 describes reduction approaches which allow to reduce the size of generated subgraphs, and sometimes entirely remove subgraphs, before decomposing them further.

## 3.1  Bounds

We consider two approaches to compute upper bounds, and two approaches to compute lower bounds, on the maximum clique size $\omega(G)$ of each subgraph $G$ generated in the graph decomposition algorithm of Sect. 2. As upper bounds we consider the following:

(i) We use a greedy search heuristic to find an upper bound on the chromatic number of the graph. The *chromatic number* of a graph $G = (V, E)$, denoted as $\chi(G)$, is the minimum number of colors needed to color each vertex of $G$ such that no edge connects two vertices of the same color. Since each vertex in a clique must have a distinct color we know that $\chi(G) \geq \omega(G)$ [17], thus the chromatic number of $G$ is an upper bound on $\omega(G)$. Computing the chromatic number is NP-hard, so its exact computation would be intractable, but there are much better heuristics for its approximation compared with the ones for the clique number. Therefore, a greedy search heuristic for the chromatic number could provide an easily computable upper bound for the maximum clique number. In the simulations of Sect. 4, we use two greedy algorithms: the greedy coloring algorithm *greedy_color* of the python module *Networkx* [16] as well as the upper bound algorithm of [4, Eq. (4)] for graphs with densities larger than 0.8.

(ii) The *Lovász number* of the complement of a graph $G$, denoted as $\theta(\overline{G})$, is defined in [18] as an upper bound on the Shannon capacity of $\overline{G}$, and is sandwiched in-between the chromatic number $\chi(G)$ and the clique number $\omega(G)$, thus $\chi(G) \geq \theta(\overline{G}) \geq \omega(G)$ [17]. Therefore, $\omega(\overline{G})$ might improve upon the bound already given by $\chi(G)$, and cannot be worse. Although the Lovász number can be calculated in polynomial time [17], existing algorithms are not very scalable (due to large constants) since they involve solving a semidefinite programming problem. The python implementation to compute the Lovász number, adapted from [27], becomes prohibitively slow for larger graphs: we only use it for subgraphs $|G| \leq 60$, and otherwise revert to the greedy search heuristic outlined above.

Likewise, we consider the following two lower bounds:

(i) Any maximum clique heuristic can be used to provide a lower bound on the clique number, provided the result returned by the heuristic is a clique. We use the *Fast Maximum Clique Finder* of [23,24] in heuristic mode.

(ii) At any point in the decomposition tree, a lower bound on one of the current subbranches can be obtained simply by solving the other subbranch. This gives a lower bound for the maximum clique size in all nodes (and leaves) of the yet unsolved branch.

It should be noted that there are many other easily computable upper and lower bounds for maximum clique, maximum independent set, and chromatic number that generally apply to any type of graph [12,26]. However, in experiments we conducted (not reported in this article) those generally applicable bounds proved to be quite conservative, with the exception of the bound of [4, Eq. (4)] which we employ for high densities.

## 3.2 Reduction Algorithms

We implement three types of reduction algorithms. The first two work directly on the subgraphs, the last one works with the QUBO formulation of MaxClique. Due to the inherent similarity between the first two techniques, we group them together as one reduction algorithm.

(i) The *(vertex) k-core algorithm* can reduce the number of vertices of the input graph in some cases, and the *edge k-core algorithm* [2,6] can reduce the number of edges.

The (vertex) $k$-core of a graph $G = (V, E)$ was defined in Sect. 2.2 as the maximal subgraph of $G$ in which every vertex has a degree of at least $k$. Therefore, if a graph has a clique $C$ of size $k + 1$, then this clique $C$ must be contained in the $k$-core of $G$ and all vertices outside of the $k$-core can be removed.

The edge $k$-core of a graph $G$ is defined in [6]. It is easily shown that for two vertices $v$, $w$ in a clique of size $c$, the intersection $N(v) \cap N(w)$ of the two neighbor lists $N(v)$ and $N(w)$ of $v$ and $w$ has size at least $c-1$. Denoting the current best lower bound on the clique size as $L$ we can therefore choose a random vertex $v$ and remove all edges $(v, e)$ satisfying $|N(v) \cap N(e)| < L-2$, since such edges cannot be part of a clique with size larger than $L$.

(ii) A second approach aims to extract information from the QUBO formulation of MaxClique in (2). For a general QUBO (1), persistency analysis of [3] allows us to identify the values that certain variables must take in any minimum (called *strong persistencies*) or in at least one minimum (called *weak persistencies*), as well as relations among them. We apply (2) to any newly generated subgraph, compute a persistency analysis for the resulting QUBO and remove a vertex $v$ from the subgraph if its corresponding variable $x_v \in \{0,1\}$ (indicating if $v$ belongs to the maximum clique or not) could be allocated a value. Furthermore, the count of solved variables which are guaranteed to be in the maximum clique for a particular subgraph can improve the lower bound on the clique number for the overall algorithm.

We use the vertex and edge $k$-cores algorithms with $k$ being set to the current best lower bound value for the clique number. This allows us to prune entire subgraphs that cannot contain a clique of size larger than the best current one. The persistency analysis allows to remove single vertices and their adjacent edges from subgraphs, as well as improve the lower bound on the clique number.

## 4   Experimental Analysis

We investigate the effectiveness of the vertex choice of Sect. 2.2 and the pruning techniques of Sect. 3, and we compare an algorithm combining both to existing work. For each graph density in the interval $[0.1, 0.9]$, incremented in steps of $0.1$, we generate random graphs of size varying from 65–110 vertices and decompose them using only the partitioning outlined in 2.1. We then generate random graphs based on each subgraph's density and size, and solve them on D-Wave. As a performance measure we report the *Time-To-Solution* measure (defined below) and/or the number of subgraphs generated, both as a function of the graph density. The Time-To-Solution (TTS) measure is defined in [20, Eq. (3)] as

$$\text{TTS} = (T_{\text{embedding}} + T_{\text{QPU}}) \frac{\log(0.01)}{\log(1-p)}, \tag{3}$$

where $T_{\text{embedding}}$ and $T_{\text{QPU}}$ are the average times in seconds (for $10^4$ anneals per graph) to compute an embedding onto the D-Wave architecture and to perform annealing, and $p$ is proportion of anneals which correctly found the maximum clique.

**Table 1.** TTS times as a function of the intended graph density (first row). The time $t_{\text{run}}$ is the sum of averages for embedding ($T_{\text{embedding}}$) and QPU solve time ($T_{\text{QPU}}$).

| Density | 0.1 | 0.2 | 0.3 | 0.4 | 0.5 | 0.6 | 0.7 | 0.8 | 0.9 |
|---|---|---|---|---|---|---|---|---|---|
| p | 0.007 | 0.003 | 0.008 | 0.015 | 0.020 | 0.003 | 0.003 | 0.007 | 0.012 |
| $t_{\text{run}}$ | 1.938 | 2.057 | 2.317 | 3.382 | 6.736 | 6.981 | 4.975 | 4.333 | 3.429 |
| TTS | 0.127 | 0.296 | 0.141 | 0.105 | 0.152 | 0.948 | 0.746 | 0.273 | 0.137 |

Table 1 shows the proportion of anneals which found the maximum clique on D-Wave, the average times $t_{\mathrm{run}}$ and the resulting TTS times for each of the graph densities considered. We fix the TTS times given in Table 1 and use them to report runtimes in the remainder of Sect. 4.

As test data, we use instances of $G(n, p)$ random graphs [13] generated with density $p$ and fixed graph size $n = 80$, which the exception of Fig. 2 which uses selected DIMACS benchmark graphs [1]. All results are averaged over 20 repetitions unless otherwise stated.

To calculate TTS for each DIMACS graph tested, we decomposed the graph being considered until we had less than or equal to 500 subgraphs. We then randomly selected 60 of those subgraphs as representatives, determined their sizes and densities, and for each we solved a random subgraph of similar size and density on D-Wave. We then applied Eq. 3 to the resulting data.

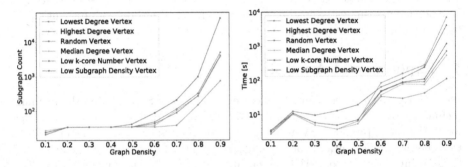

**Fig. 2.** Subgraph count (left) and total solution time (right) as a function of the graph density for the six vertex selection approaches of Sect. 2.2. Log scale on the y-axes.

### 4.1  Vertex Choice

Figure 2 shows the total solution time to decompose our test graphs (to the point that each subproblem reaches a size of at most 46 vertices, thus making it possible to compute the maximum clique on D-Wave) as a function of the graph density. We compare the six different vertex selection choices delineated in Sect. 2.2. The lowest degree selection strategy seems to lead to the best solution time among the six approaches, where the advantage of the lowest degree selection becomes more pronounced as the graph density increases. Similar results are observed for the number of subgraphs.

### 4.2  Bounds

Figure 3 (left) shows the number of subgraphs generated by every combination of the two lower and upper bounds outlined in Sect. 3.1. Figure 3 (right) shows the speedup in runtime when using the upper bounds of Sect. 3.1 compared to not using any bounds. The speedup is defined as $T_0/T_1$, where $T_0$ ($T_1$) is the total solution time without (with) bounds. The right figure shows that the Lovász number bound is not an efficient bound in terms of solution time due to the preprocessing CPU time required to calculate it. However, the Lovász number leads to the lowest number of generated subgraphs (left figure). Over most densities tested, the combination of fast heuristic bounds given by upper bound (i) and lower bound (ii) (see Sect. 3.1) seem to give a reasonable trade-off between total solution time and subgraph count. We will use this choice in Sect. 4.4.

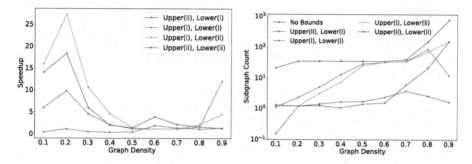

**Fig. 3.** Subgraph count (left) and ratio $T_0/T_1$ (right) defined in Sect. 4.2 as a function of the graph density for the two techniques to compute upper bounds on the clique number (see Sect. 3.1), in comparison to no bound calculations. Log scale on the y-axis showing subgraph count.

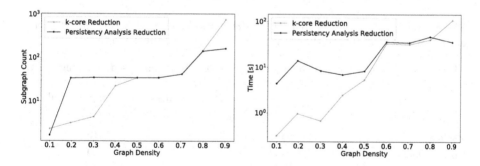

**Fig. 4.** Subgraph count (left) and total solution time (right) for the two reduction techniques from Sect. 3.2. Log scale on the y-axes.

### 4.3 Reduction Algorithms

Figure 4 (right) shows the total runtime to decompose our test graphs as a function of the density $p$. Two methods are compared: the $k$-core reduction, and the persistency analysis (Sect. 3.2). Figure 4 (left) shows that the $k$-core reduction algorithm can significantly reduce the number of subgraphs, and therefore the total solution time, over more densities than the persistency analysis can, and that the latter yields improvements in runtime only for high densities with $p > 0.8$ (Fig. 4, right).

## 4.4    Comparisons

We denote the decomposition algorithm of Sect. 2 which uses the choice of fast heuristic bounds of Sect. 3.1 for pruning subgraphs and the $k$-core of Sect. 3.2 as the DBK algorithm (Decomposition, Bounds, $k$-core). Figure 5 shows the speedup of the DBK algorithm over the decomposition algorithm presented in [6]. To be precise, Fig. 5 displays $T_{\text{Chapuis}}/T_{\text{DBK}}$, where $T_{\text{DBK}}$ is the time the DBK algorithm takes for a particular graph instance, and analogously for $T_{\text{Chapuis}}$. The figure shows that both algorithms are quite similar for graph densities in the interval $[0.4, 0.8]$, however DBK considerably improves upon [6] the sparser or denser the graphs are.

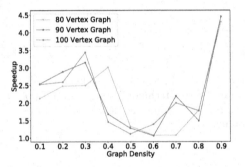

**Fig. 5.** Ratio $T_{\text{Chapuis}}/T_{\text{DBK}}$ defined in Sect. 4.4 showing the speedup for the DBK algorithm in comparison to the algorithm of [6] for low and high densities.

**Table 2.** Total solution time in seconds for DIMACS graphs based on a single run.

| DIMACS graph | No. vertices | No. edges | $\omega(G)$ | DBK algorithm | Chapuis et al. [6] |
|---|---|---|---|---|---|
| johnson16-2-4 | 120 | 5460 | 8 | 3217 | 3217 |
| keller4 | 171 | 9435 | 11 | 682 | 1310 |
| p-hat300-1 | 300 | 10933 | 8 | 16 | 20 |
| p-hat300-2 | 300 | 21928 | 25 | 794 | 2781 |
| p-hat500-1 | 500 | 4459 | 9 | 61 | 77 |
| p-hat700-1 | 700 | 60999 | 11 | 175 | 643 |
| brock200-2 | 200 | 9876 | 12 | 72 | 295 |
| brock200-3 | 200 | 12048 | 15 | 1185 | 1640 |
| brock200-4 | 200 | 13089 | 17 | 2168 | 14735 |
| hamming6-2 | 64 | 1824 | 32 | 4 | 17 |
| hamming8-4 | 256 | 20864 | 16 | 5179 | 18253 |

For DIMACS graphs displayed in Table 2, we observe that the DBK algorithm can yield substantial improvements in runtime especially for large graphs, and that it never performs worse than [6].

## 5   Discussion

This article investigates subgraph extraction and pruning techniques for a decomposition algorithm designed to compute maximum cliques of a graph. These reduction techniques aim at reducing the number of recursively generated subgraphs for which the maximum clique problem needs to be solved. The overall runtime of our algorithm depends on the user's choice of method to solve the decomposed maximum clique subproblems at leaf level of the decomposition. Our analysis shows the following:

1. The decomposition is sensitive to the vertex choice used to extract subgraphs. Extraction of the subgraph induced by the lowest degree vertex seems to lead to the lowest overall runtime.
2. A variety of lower and upper bounds exist to estimate the maximal clique size and help prune subgraphs which cannot contain the maximum clique. Of those investigated, the Lovász bound is most effective in pruning subgraphs but also most computationally intensive, whereas a heuristic upper bound leads to a good trade-off between subgraph reduction and runtime.
3. The $k$-core reduction prunes subgraphs more effectively than an alternative approach based on persistency analysis.
4. Combining the decomposition algorithm with subgraph extraction and fast heuristic bounds on the clique size for pruning leads to an improved algorithm which, despite the additional computations in each recursion, is faster at computing maximum cliques than a previous algorithm of [6], with a considerable speed-up for very sparse and very dense graphs.

Future research could investigate the viability of trade-off algorithms, where differing decomposition and simplification strategies are used in response to different problem graph types. Another future improvement concerns the Lovász number: our experiments show that its bounds yield the best reduction in subproblem count, but also come at the highest computational cost (see Fig. 3). Additionally, the impracticality of applying the adapted python implementation [27] on large graphs was not ideal. A faster way to compute the Lovász number would render those bounds usable in practice.

**Acknowledgments.** Research presented in this article was supported by the Laboratory Directed Research and Development program of Los Alamos National Laboratory under project number 20180267ER.

# References

1. Bader, D.A., Meyerhenke, H., Sanders, P., Wagner, D.: Graph partitioning and graph clustering. In: 10th DIMACS Implementation Challenge Workshop, 13–14 February 2012. Contemporary Mathematics, vol. 588 (2013)
2. Batagelj, V., Zaversnik, M.: An O(m) algorithm for cores decomposition of networks. Adv. Dat An Class **5**(2) (2011)
3. Boros, E., Hammer, P.: Pseudo-Boolean optimization. Discret. Appl. Math. **123**(1–3), 155–225 (2002)
4. Budinich, M.: Exact bounds on the order of the maximum clique of a graph. Discret. Appl. Math. **127**(3), 535–543 (2003)
5. Carmo, R., Züge, A.: Branch and bound algorithms for the maximum clique problem under a unified framework. J. Braz. Comput. Soc. **18**(2), 137–151 (2012)
6. Chapuis, G., Djidjev, H., Hahn, G., Rizk, G.: Finding maximum cliques on the D-wave quantum annealer. In: Proceedings of the 2017 ACM International Conference on Computing Frontiers (CF 2017), pp. 1–8 (2017)
7. Courcelle, B., Makowsky, J., Rotics, U.: Linear time solvable optimization problems on graphs of bounded clique-width. Theor. Comput. Syst. **33**(2), 125–150 (2000)
8. D-Wave: Technical Description of the D-Wave Quantum Processing Unit, 09-1109A-A, 2016 (2016)
9. D-Wave Systems (2000). Quantum Computing for the Real World Today
10. Dabrowski, K., Lozin, V., Müller, H., Rautenbach, D.: Parameterized algorithms for the independent set problem in some hereditary graph classes. In: Iliopoulos, C.S., Smyth, W.F. (eds.) IWOCA 2010. LNCS, vol. 6460, pp. 1–9. Springer, Heidelberg (2011). https://doi.org/10.1007/978-3-642-19222-7_1
11. Djidjev, H., Hahn, G., Niklasson, A., Sardeshmukh, V.: Graph partitioning methods for fast parallel quantum molecular dynamics. In: SIAM Workshop on Combinatorial Scientific Computing CSC 2016 (2015)
12. Elphick, C., Wocjan, P.: Conjectured lower bound for the clique number of a graph. arXiv:1804.03752 (2018)
13. Erdös, P., Rényi, A.: On the evolution of random graphs. Publ. Math. Inst. Hungarian Acad. Sci. **5**, 17–61 (1960)
14. Giakoumakis, V., Vanherpe, J.: On extended P4-reducible and extended P4-sparse graphs. Theor. Comput. Sci. **180**, 269–286 (1997)
15. Grötschel, M., Lovász, L., Schrijver, A.: Geometric Algorithms and Combinatorial Optimization. Springer, Berlin (1988). https://doi.org/10.1007/978-3-642-78240-4
16. Hagberg, A., Schult, D., Swart, P.: Exploring network structure, dynamics, and function using NetworkX. In: Proceedings of SciPy 2008, pp. 11–15 (2008)
17. Knuth, D.E.: The sandwich theorem. Electron. J. Comb. **1**(A1), 1–49 (1993)
18. Lovász, L.: On the Shannon capacity of a graph. IEEE Trans. Inf. Theory **IT–25**(1), 1–7 (1979)
19. Lucas, A.: Ising formulations of many NP problems. Front. Phys. **2**(5), 1–27 (2014)
20. Mandrà, S., Katzgraber, H.: A deceptive step towards quantum speedup detection. Quant. Sci. Technol. **3**(04LT01), 1–12 (2018)
21. Minty, G.: On maximal independent sets of vertices in claw-free graphs. J. Comb. Theory Ser. B **28**(3), 284–304 (1980)
22. Pardalos, P.M., Rodgers, G.P.: A branch and bound algorithm for the maximum clique problem. Comput. Oper. Res. **19**(5), 363–375 (1992)
23. Pattabiraman, B., Patwary, M., Gebremedhin, A., Liao, W.K., Choudhary, A.: Fast max-clique finder (2018). http://cucis.ece.northwestern.edu/projects/MAXCLIQUE

24. Pattabiraman, B., Patwary, M., Gebremedhin, A., Liao, W.K., Choudhary, A.: Fast algorithms for the maximum clique problem on massive graphs with applications. Internet Math. **11**, 421–448 (2015)
25. Rao, M.: Solving some NP-complete problems using split decomposition. Discret. Appl. Math. **156**(14), 2768–2780 (2008)
26. Soto, M., Rossi, A., Sevaux, M.: Three new upper bounds on the chromatic number. Discret. Appl. Math. **159**, 2281–89 (2011)
27. Stahlke, D.: Python code to compute the Lovasz, Schrijver, and Szegedy numbers for graphs (2013). https://gist.github.com/dstahlke/6895643
28. Tarjan, R.: Decomposition by clique separators. Discret. Math. **55**(2), 221–232 (1985)

# Quantum Annealing Based Optimization of Robotic Movement in Manufacturing

Arpit Mehta[✉], Murad Muradi, and Selam Woldetsadick

BMW AG, Munich, Germany
{Arpit.Mehta,Murad.Muradi,Selam-Getachew.Woldetsadick}@bmwgroup.com

**Abstract.** Recently, considerable attention has been paid to planning and scheduling problems for multiple robot systems (MRS). Such attention has resulted in a wide range of techniques being developed in solving more complex tasks at ever increasing speeds. At the same time, however, the complexity of such tasks has increased as such systems have to cope with ever increasing business requirements, rendering the above mentioned techniques unreliable, if not obsolete. Quantum computing is an alternative form of computation that holds a lot of potential for providing some advantages over classical computing for solving certain kinds of difficult optimization problems in the coming years. Motivated by this fact, in this paper we demonstrate the feasibility of running a particular type of optimization problem on existing quantum computing technology. The optimization problem investigate arises when considering how to optimize a robotic assembly line, which is one of the keys to success in the manufacturing domain. A small improvement in the efficiency of such an MRS can lead to huge saving in terms of time of manufacturing, capacity, robot life, and material usage. The nature of the quantum processor used in this study imposes the constraint that the optimization problem be cast as a quadratic unconstrained binary optimization (QUBO) problem. For the specific problem we investigate, this allows situations with one robot to be modeled naturally, meanwhile modeling the multi-robot generalization is less obvious and left as a topic for future research. The results show that for simple 1-robot tasks, the optimization problem can be straightforwardly solved within a feasible time span on existing quantum computing hardware.

**Keywords:** Quantum annealing · Robotic manufacturing · TSP · MRS · D-Wave · QUBO · Optimization problem

## 1 Introduction

Quantum computing technology is widely considered to constitute a promising platform for providing some advantages over classical computing in the coming years for solving certain kinds of difficult optimization problems, such as problems arising in combinatorial optimization. An example of a combinatorial optimization problem is as follows. From a large finite set of possible production

© Springer Nature Switzerland AG 2019
S. Feld and C. Linnhoff-Popien (Eds.): QTOP 2019, LNCS 11413, pp. 136–144, 2019.
https://doi.org/10.1007/978-3-030-14082-3_12

plans, the specific subset must be found which optimizes the desired target functions while satisfying predetermined constraints. Due to the high computational effort, exact optimization can often take infeasibly long for large enough problems. In practice, one therefore typically resorts to specialized heuristics which often times have no formal guarantee of returning a global optimum for short run times, but have nevertheless been shown experimentally to find, in a feasible period of time, solutions which are "good enough" for the specific application at hand. One may view quantum annealing as one such heuristic, and in this paper we investigate its performance on a specific optimization problem.

The problem we investigate is similar to the Traveling Salesman Problem (TSP), which is ubiquitous across many areas of optimization and is one of the most intensely studied problems in optimization. We begin with an informal description of the problem we investigate. Suppose there are $N$ 1-dimensional paths. For now, most of the details of each path are irrelevant to us: of relevance to us are only the start- and end-points of each path. Each path corresponds to a task on which a robot, with a given starting location, needs to do some work. There are, of course, several natural extensions of the problem that one might consider for a more realistic treatment of a real-world problem, such as, for example, the generalization to $M \geq 1$ robots. In this paper, however, we focus on the simple case of 1 robot. In the automotive industry, there are numerous robot applications that correspond to the described problem formulation, such as laser welding, seam sealing, measuring, gluing and painting. Informally speaking, the relevant computational problem is the problem of determining in which order the tasks should be done so as to minimize the total length traveled by the robot, thus minimizing the time required to perform the tasks.

In Fig. 1a typical assembly line motion is illustrated. A typical process requires firstly the distribution of tasks among the robots and secondly scheduling the order of tasks for each robot. As seen in Fig. 1 firstly distributing tasks among robot 1, 2 and 3 is an optimization problem, secondly finding an optimal order to complete the allocated tasks is also an optimization problem. For example, Robot 1 starts from it's initial position, completes task b, followed by task c and ends back at its initial position.

In section the following section, the exact problem we study is formally and mathematically described.

## 2    Formal Description of the Problem

### 2.1    Modelling the Problem with a Weighted Graph

The problem of interest can be modeled by a weighted graph, which is denoted as $G := (V, E, W)$. The vertices correspond to the starting locations of the robot, $v_r$, and the start- and endpoints of each task. We can without loss of generality arbitrarily set one in each of those pairs of points to be the starting point, and denote those $V^{(1)} := \{v_1^{(1)}, \ldots, v_N^{(1)}\}$; similarly, we denote the endpoints $V^{(2)} := \{v_1^{(2)}, \ldots, v_N^{(2)}\}$. The vertices of $G$ are then given as follows:

$$V := \{v_r\} \cup V^{(1)} \cup V^{(2)}.$$

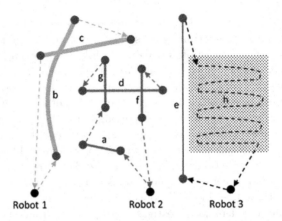

**Fig. 1.** Simplified illustration of the optimization problem in 2D.

The edges of $G$ correspond to the possible paths that the robot can travel. These edges take one of two forms. First, they can be of the form $(v_r, v_i^{(k)})$, where $i \in \{1, \ldots, N\}, k \in \{1, 2\}$. Edges of this form correspond to paths between the starting location of the robot, and the start- or endpoint of a task. Alternatively, edges can be of the form $(v_i^{(k)}, v_j^{(l)})$. Edges of this form correspond to the paths between either the start- or endpoint of task $i$, to either the start- or endpoint of task $j$. The union of all such pairs of vertices forms the set of edges of $G$:

$$E := (\cup_{i=1}^N \cup_{k=1}^2 \{(v_r, v_i^{(k)})\}) \cup (\cup_{i,j=1}^N \cup_{k,l=1}^2 \{(v_i^{(k)}, v_j^{(l)})\}).$$

Finally, the weight of each edge will correspond to the "cost" of traversing each edge. In the simplest case, this can correspond to the distance between two vertices. However, one can also imagine incorporating other relevant features into the weight function. For instance, if the robot has to change it's mounted tool between a pair of tasks, effectively delaying the time it takes for it to be able to get started processing the second task in this pair, one possible way of modeling this is by increasing the weight of the corresponding edge. In this paper, however, we focus on the simpler case where the weight of each edge corresponds to its length:

$$W : E \to \mathbb{R},$$

$$W(\{v_1, v_2\}) = d(v_1, v_2).$$

Now that we have constructed the graph $G = (V, E, W)$, we are ready to define the QUBO that models our problem. Recall that if the goal is to solve the problem on a D-Wave quantum annealer, our objective function must be a quadratic function, i.e., any function of the form $f(x) := x^T Q x$. As we will see, one can do this naturally in the case of 1 robot. Following the work from [4], we proceed as follows.

## 2.2   The Logical Variables

We first describe the logical variables $\boldsymbol{x}$. For each $i, t \in \{1, \ldots, N\}, d \in \{1, 2\}$, we associate a binary variable $x_{i,t}^{(d)} \in \{0, 1\}$, which encodes whether or not the $t$-th task to perform should be task $i$ in direction $d$; without loss of generality, we arbitrarily define direction $d$ to be the direction such that when a robot performs task $i$, the robot ends at vertex $v_i^{(d)}$. We put all of these variables together in one vector and obtain the binary vector $\boldsymbol{x} \in \{0, 1\}^{2N^2}$, i.e.:

$$
\begin{aligned}
\boldsymbol{x} = (&x_{1,1}^{(1)}, x_{1,1}^{(2)}, x_{2,1}^{(1)}, x_{2,1}^{(2)}, \ldots, x_{N,1}^{(1)}, x_{N,1}^{(2)}, \\
&x_{1,2}^{(1)}, x_{1,2}^{(2)}, x_{2,2}^{(1)}, x_{2,2}^{(2)}, \ldots, x_{N,2}^{(1)}, x_{N,2}^{(2)}, \\
&\vdots \\
&x_{1,N}^{(1)}, x_{1,N}^{(2)}, x_{2,N}^{(1)}, x_{2,N}^{(2)}, \ldots, x_{N,N}^{(1)}, x_{N,N}^{(2)}).
\end{aligned}
$$

Informally, $\boldsymbol{x}$ encodes the order in which the robot is to perform the tasks.

## 2.3   The Objective Function

Now that we have defined the logical variables, we can define the objective function. For now we focus on an objective function which will model the distance traveled by the robot between tasks. This is given by

$$
\begin{aligned}
f_{\text{distance}} = &\sum_{i=1}^{N} \sum_{k=1}^{2} W(\{v_r, v_i^{(k')}\}) x_{i,1}^{(k)} \\
&+ \sum_{t=1}^{N} \sum_{i,j=1}^{N} \sum_{k,l=1}^{2} W(\{v_i^{(k)}, v_j^{(l)}\}) x_{i,t}^{(k)} x_{j,t+1}^{(l')},
\end{aligned} \tag{1}
$$

where $v_r$ is the vertex corresponding to the robot, and $\{k, k'\} = \{l, l'\} = \{1, 2\}$. Informally, the first term in Eq. (1) corresponds to the distance the robot travels to its first task, and the second term corresponds to the distance the robot travels between subsequent tasks. Note that from a purely formal point of view, $f_{\text{distance}}$ is minimized for $\boldsymbol{x} = 0$, which corresponds to the assignment in which no tasks are performed. Clearly this is not a physically relevant solution. Therefore, in addition to $f_{\text{distance}}$, the final objective function needs to include penalty functions which penalize such physically irrelevant assignments. In particular, as we will see, we require two penalties.

   The first penalty enforces the constraint that all tasks be performed *exactly* once. For each task $i$, this corresponds the equality constraint

$$
\sum_{t=1}^{N} \sum_{d=1}^{2} x_{i,t}^{(d)} = 1. \tag{2}
$$

A quadratic function which penalizes values of $x$ which don't satisfy this equality is given by

$$f_{\text{tasks}}^{(i)} := P_i \left( \sum_{t=1}^{N} \sum_{d=1}^{2} x_{i,t}^{(d)} - 1 \right)^2, \tag{3}$$

where $P_i \in (0, \infty)$ must be chosen large enough so that the penalty is enforced. While one might be tempted to set $P_i$ to some arbitrarily large number, one must take into account the limited precision available on current quantum computing hardware. Setting $P_i$ too large may result in the quantum processor not being able to accurately implement the rest of the problem. Therefore, in practice one strives to choose the smallest possible $P_i$ that still effectively enforces these constraints. With this in mind, in our experiments we set $P_i$ to the following:

$$P_i = \max_{\substack{j \in \{1,\dots,N\} \\ l,k \in \{1,2\}}} \max\{W(\{v_j^{(l)}, v_i^{(k)}\}), W(\{v_r, v_i^{(k)}\})\}$$

$$+ \max_{\substack{j \in \{1,\dots,N\} \\ k,l \in \{1,2\}}} W(\{v_i^{(k)}, v_j^{(l)}\}). \tag{4}$$

Intuitively, $P_i$ is equal to the sum of the largest distance that can be traveled *to* task $i$ (first term in the sum) and the largest distance that can be traveled *from* task $i$ (second term in the sum). Summing over all tasks, we obtain the penalty function which enforces that all tasks be performed exactly once:

$$f_{\text{tasks}} := \sum_{i=1}^{N} f_{\text{tasks}}^{(i)}. \tag{5}$$

Note that from a purely formal point of view, the assignment $x_{i,2}^{(1)} = 1$ for all $i$, and $x_{j,t}^{(d)} = 0$ otherwise minimizes both $f_{\text{distance}}$ and $f_{\text{tasks}}$. This corresponds to an assignment for which the robot takes care of all tasks *simultaneously*. Clearly this is not a physically relevant solution, since the robot can do at most one task at a time. This leads us to the second and final penalty function we require.

The second penalty should enforce the constraint that at each time step, *exactly* one task be performed. For each time step $t$, this corresponds to the equality constraint

$$\sum_{i=1}^{N} \sum_{d=1}^{2} x_{i,t}^{(d)} = 1. \tag{6}$$

A quadratic function which penalizes values of $x$ which don't satisfy this equality is given by

$$f_{\text{time}}^{(t)} := P_t \left( \sum_{i=1}^{N} \sum_{d=1}^{2} x_{i,t}^{(d)} - 1 \right)^2, \tag{7}$$

where $P_t \in (0, \infty)$ must be chosen large enough so that the penalty is enforced. In practice, similar considerations must be taken into account when choosing $P_t$ as when choosing $P_i$. In our experiments, we set $P_t$ to the following:

$$P_t = \max_{i \in \{1,\dots,N\}} P_i. \tag{8}$$

Intuitively, $P_t$ is the largest possible distance that the robot can travel at any time step. This is in turn equal to the sum of the largest distance that can be traveled to any task plus the largest distance that can be traveled from that task. Summing over all time steps, we obtain the penalty function which enforces that exactly one task be performed per time step:

$$f_{\text{time}} := \sum_{t=1}^{N} f_{\text{time}}^{(t)}. \tag{9}$$

Summing together Eqs. (1), (5), and (9), we obtain our final objective function:

$$f : \{0,1\}^{2N^2} \to \mathbb{R} \tag{10}$$

$$f := f_{\text{distance}} + f_{\text{tasks}} + f_{\text{time}}. \tag{11}$$

## 3   Results

Based on the problem formulation in the previous section, this section provides results of the experiment. For each $N \in \{2,3,4,5\}$, we generated 20 instances with $N$ tasks, with the location of the end points of each task chosen at random. Note that once the vertices are determined, the edges and the corresponding edge weights are defined according to the formulation described in the previous section. In particular, given a set of vertices, we obtain a function $f$ of the form in Eq. 11. It is this function that we then solved with the D-Wave 2000Q machine [3].

### 3.1   Quantum Annealing vs. Classical Method

The results of the quantum annealer were compared with those of a classical method. The classic algorithm chosen was the Herd-Karp algorithm [2]. It is an exact method for solving TSP and has a exponential time complexity of $O(2^n n^2)$. Table 1 lists the start and end positions of each task along with the associated speed. The processing time of a task results from the Euclidean distance between start and end position divided by the velocity. The travel time between two separate tasks is calculated by the corresponding Euclidean distance of the link divided by a default speed of 100. The robot home position (R0) has been set to (0, 0).

Due to limitation on the D-Wave hardware the problem size for this experiment was chosen to be minimal. The optimal solution of 52.08 s – verified by the Held-Karp algorithm – was also found by the quantum annealer.

**Table 1.** The tasks information for the problem instance with size 5.

| Task | Start | End | Velocity |
|------|-------------|-------------|----------|
| a | (287, 619) | (17, 479) | 94 |
| b | (595, 627) | (592, 52) | 58 |
| c | (488, 353) | (43, 565) | 68 |
| d | (450, 142) | (688, 580) | 63 |
| e | (136, 403) | (630, 170) | 70 |

The optimal path is: R0 → a' → c' → d → b → e' → R0.

Here ' indicates that the task is processed in the reverse direction.

The results supported the hypothesis that quantum computing more specifically annealing approach is a viable alternative for solving optimization problems. Hence, future developments in quantum computer capability increase potential of solving real business challenges.

## 3.2   Performance Testing of the QC Algorithm

The results of performance testing are shown in Fig. 2. The figure shows the success probability for any given annealing cycle on the D-Wave quantum annealer versus problem size. For small problems with two tasks, the observed median success probability per annealing cycle was ≈0.3. By *success* we mean that the annealing cycle returned a configuration that resulted in the optimal solution *after* repairing every so-called *chains*, which is a group of qubits corresponding to the same logical variable. For the largest problem sizes investigated, namely 5 tasks, the observed median success probability per annealing cycle was ≈$10^{-5}$. Also shown in Fig. 2 are approximate corresponding wall-clock times for each success probability. These wall-clock times are computed as follows. Each *call* to the D-Wave quantum annealer is restricted to at most $10^4$ annealing cycles. Furthermore, each call takes ≈5 s of wall-clock time to complete. This means that one can obtain an approximate estimate of the wall clock-time $t_{clock}$ corresponding to a success probability $p$ via the equation $t_{clock} = \lceil \frac{1/p}{10^4} \rceil \times 5\,\text{s}$.

For the range of problem sizes investigated, the median success probability appears to decay exponentially with the number of tasks. Note that for such a combinatorial optimization problem, this is the kind of scaling one would expect with any *exact* solver, whether quantum or classical. Relatedly, it is important to note that these results indicate a sort of "worst-case scenario" performance. More concretely, (i) no preprocessing was done on the QUBOs before being submitted to the DW quantum annealer, and (ii) the success probabilities indicate the probability of finding a *global* optimum. In practice, (i) state-of-the-art techniques will employ heuristics to speed up the computation time, and (ii) typically it is only necessary to find a "good enough" solution, as opposed a global optimum. We leave it as an interesting avenue of future research to see how the results might change if we take into account points (i) and (ii).

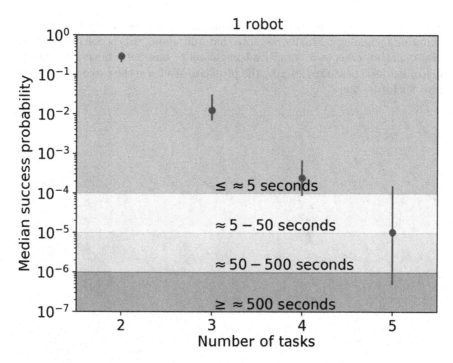

**Fig. 2.** Median success probabilities. Error bars correspond to lower and upper quartiles. Time estimates denote approximate wall-clock times corresponding to each success probability.

## 4 Conclusion

In this paper we investigated a simple prototype implementation of a simple TSP-like problem arising in MRS. We emphasize that the goal of this simple study is to show the feasibility of running a particular kind of optimization problem on a quantum annealer.

The question of how one might do this in such a way that a computational advantage is obtained lies far outside the scope of this project. It would be interesting to investigate how one might implement more sophisticated models. In addition, it would be interesting to investigate how one might implement heuristics to improve the efficacy of the quantum annealer *in practice*. For a glimpse of how some approaches might look like, consider the fact that one could use several heuristic approaches to reducing the $M$-robot setting to the 1-robot setting, and then using the algorithm developed in detail on each of the derived 1-robot problems. A second line of investigation that would be interesting to investigate would be to first apply heuristics to create a population of good solutions and then use AI to identify a common pattern in it. In this way, it is possible to reduce the number of logical variables by considering frequently occurring task pairs as a fixed sequence (see Fig. 3). We anticipate should heuristics would allow one to address larger and overall more complex problems on existing quantum computing technology. Finally, we note that addressing this problem with a universal quantum computer could lead might allow one more flexibility with the kinds of models chosen to describe the problem. We leave the exploration of such topics for future work.

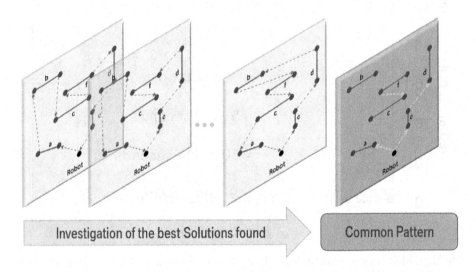

**Fig. 3.** Useage of heuristics and AI to reduce the number of logical variables.

# References

1. Correl, R.: Quantum artificial intelligence and machine learning: the path to enterprise deployments. QCWare (2017). https://dwavefederal.com/app/uploads/2017/10/QC-Ware-Qubits-20170926-Final.pdf
2. Held, M., Karp, R.M.: The traveling-salesman problem and minimum spanning trees. Oper. Res. **18**(6), 1138–1162 (1970)
3. D.W.S, Inc.: The D-wave 2000Q quantum computer technology overview. D-Wave (2018)
4. Lucas, A.: Ising formulations of many NP problems. Front. Phys. **2**, 5 (2014). https://doi.org/10.3389/fphy.2014.00005. https://doi.org/10.3389/fphy.2014.00005
5. Morita, S., Nishimori, H.: Convergence theorems for quantum annealing. J. Phys. A: Math. Gen. **39**(45), 13903 (2006). http://stacks.iop.org/0305-4470/39/i=45/a=004
6. Moylett, D.J., Linden, N., Montanaro, A.: Quantum speedup of the traveling-salesman problem for bounded-degree graphs. Phys. Rev. A **95**, 032323 (2017). https://doi.org/10.1103/PhysRevA.95.032323

# Quantum Annealing of Vehicle Routing Problem with Time, State and Capacity

Hirotaka Irie[1]([⊠]), Goragot Wongpaisarnsin[2], Masayoshi Terabe[1], Akira Miki[1], and Shinichirou Taguchi[1]

[1] Information Electronics R&I Department, Electronics R&I Division, DENSO Corporation, Tokyo Office, Tokyo 103-6015, Japan
{hirotaka_irie,masayoshi_i_terabe,akira_miki, shinichirou_taguchi}@denso.co.jp
[2] Contents Development & Distribution Department, Toyota Tsusho Nexty Electronics (Thailand) Co., LTD., Bangkok 10330, Thailand
goragot@th.nexty-ele.com

**Abstract.** We propose a brand-new formulation for capacitated vehicle routing problem (CVRP) as a quadratic unconstrained binary optimization (QUBO). The formulated CVRP is equipped with *time-table* which describes time-evolution of each vehicle. Therefore, various constraints associated with time can be successfully realized. Similarly, constraints of *capacities* are also introduced, where capacitated quantities are allowed to increase and decrease according to the cities in which vehicles arrive. As a bonus of capacity-qubits, one can also obtain a description of *state*, which allows us to set various traveling rules, depending on the state of each vehicle. As a consistency check, the proposed QUBO formulation is also evaluated using a quantum annealing machine, D-Wave 2000Q.

**Keywords:** Vehicle routing problem · QUBO · Quantum annealing

## 1 Introduction

Vehicle Routing Problem (VRP) [1] is a basic mathematical problem related to optimization of planning, logistics, and transportation. Owing to the recent interests in quantum annealing machines, which were first studied theoretically by Kadowaki-Nishimori [2] and made available commercially by D-Wave Systems Inc. [3], the investigation of VRP as a quadratic unconstrained binary optimization (QUBO) has become very important, particularly in an attempt to achieve quantum-mechanical optimization of real-world problems encountered in our daily life concerning various mobility services.

Based on a joint project of DENSO Corporation, Toyota Tsusho Corporation and Toyota Tsusho Nexty Electronics (Thailand) Co., LTD. The authors greatly thank Shunsuke Takahashi, Masakazu Gomi and Toru Awashima for valuable discussions which make this work possible.

S. Feld and C. Linnhoff-Popien (Eds.): QTOP 2019, LNCS 11413, pp. 145–156, 2019.
https://doi.org/10.1007/978-3-030-14082-3_13

VRP is a generalization of Traveling Salesman Problem (TSP), i.e., a problem to find the traveling path that has the lowest cost. Similarly, the purpose of VRP is to find the best routing scheme for multiple vehicles that achieves the lowest cost under various circumstances; for instance, each city should be visited exactly once. QUBO formulation of the TSP is found in [4]; it was constructed by adding the square of linear-constraint functions to the associated cost functions [5–7] using Lagrange-like multipliers. A straightforward extension of this QUBO formulation can also be applied to VRP by introducing several copies of QUBO systems for TSP. However, such formulation intrinsically suffers from the strict concept of *time*. It does have a concept of time-step, but it is generally not equivalent to the concept of time in many practical applications. Obviously, if one of the cities is located far from the rest, then traveling there will take more time than traveling to other cities. In such case, the conventional time-step formulation cannot describe the time, which flows commonly and homogeneously for all vehicles. Therefore, the introduction of time in the conventional QUBO formulations is the main obstacle in formulating various important VRP constraints associated with time such as time-window, (non-)simultaneous arrivals, and chronological variation of cost.

In addition to the concept of time, it is also important to introduce the concept of *capacity* (i.e., capacitated VRP or CVRP), which may describe the capacity of carrying passengers or packages. There are some attempts toward a QUBO formulation of CVRP: one tackles a possibility of constraint terms describing inequality [8]; and another utilizes a hybrid cluster algorithm combined with the TSP QUBO systems [9]. In this paper, we propose a different approach to investigate a concise QUBO formulation. In fact, capacity and time are similar as the description of time already implies that vehicles should travel within their own capacity of time. As further explained, these two concepts are similarly implemented by introducing a new kind of interactions that depend on the cities of departure and destination only. Furthermore, by introducing *capacity-qubits* (See Sect. 3), one can realize multiple capacitated variables which can be allowed to increase and decrease (i.e. pickup and delivery) during each travel.

However, there are further applications of this approach. As a bonus of capacity-qubits, one can also describe the concept of *state*. In particular, cost and time-duration can change depending on each state of vehicles. As a simplest example, we demonstrate a two-state model: arrival-state and departure-state. The mean time from arrival to departure is the duration of visits. We can also set up how long each vehicle stays depending on the cities visited by vehicles.

The organization of this paper is as follows: The brand-new QUBO formulation of VRP is introduced with time-table (in Sect. 2), with capacity-qubits (in Sect. 3), and with the concept of state (in Sect. 4). Other related constraints from real-world applications are in Sect. 5, and validity of our models is discussed in Sect. 6. Conclusion and discussion are presented in Sect. 7.

## 2    Time-Table in TSP/VRP

The first new ingredient introduced to the TSP/VRP concept in this paper is *time*. In particular, we introduce *time-table* in the TSP/VRP formulation. First, we would prepare binary qubits parametrized by three integers $(\tau, a, i)$:

$$x_{\tau,a}^{(i)} \qquad \left(1 \leq \tau \leq T, \quad 1 \leq a \leq N, \quad 1 \leq i \leq k\right), \tag{1}$$

where $\tau$ parametrizes each *time-interval* of the time-table, with the assumption that there are $N$ cities to visit and $k$ vehicles are present. The time-interval means that we divide the total time into several time units as follows:

$$(\tau) \qquad \longmapsto \qquad [t_\tau, t_{\tau+1}] \qquad (1 \leq \tau \leq T-1), \qquad t_{\tau+1} - t_\tau \equiv \Delta t. \tag{2}$$

Herein $\Delta t$ is the unit of time-division.[1] For example, three hours from 9:00 AM to 12:00 AM can be divided into nine intervals with twenty-minute duration each. Suppose a vehicle $(i)$ does (or does not) arrive at a city $(a)$ in a time-interval $(\tau)$, the binary qubits $x_{\tau,a}^{(i)}$ takes the following value:

$$x_{\tau,a}^{(i)} = 1 \quad \text{(arriving)}, \qquad x_{\tau,a}^{(i)} = 0 \quad \text{(not arriving)}. \tag{3}$$

Since the conventional QUBO formulation of TSP/VRP always assumes one-step forward, we would invent a new kind of interaction which describes how much time each vehicle spends for each travel. Thus, we would first introduce *time-duration matrices* $\left(n_{ab}^{(\tau)}\right)_{1 \leq a \neq b \leq N}$ as well as cost matrices $\left(d_{ab}^{(\tau)}\right)_{1 \leq a \neq b \leq N}$ for each time-interval $(\tau)$ as[2]

$$n_{ab}^{(\tau)} = \left\lceil \frac{\left(\text{time spent from a city } (b) \text{ (at time } \tau) \text{ to a city } (a)\right)}{\Delta t} \right\rceil \geq 1, \tag{4}$$

$$d_{ab}^{(\tau)} = \left(\text{cost spent from a city } (b) \text{ (at time } \tau) \text{ to a city } (a)\right). \tag{5}$$

Thus, our proposed Hamiltonian $\mathcal{H}$ can then be written as follows:

$$\mathcal{H} = \sum_{\left\{\substack{1 \leq a \neq b \leq N \\ 1 \leq \tau \leq T-1 \\ 1 \leq i \leq k}\right\}} \left( \underbrace{\frac{d_{ab}^{(\tau)} - \mu}{\rho} \times x_{\tau+n_{ab}^{(\tau)},a}^{(i)} \, x_{\tau,b}^{(i)}}_{(*1)} + \sum_{1 \leq \delta\tau \leq n_{ab}^{(\tau)}-1} \underbrace{\lambda \times x_{\tau+\delta\tau,a}^{(i)} \, x_{\tau,b}^{(i)}}_{(*2)} \right)$$

$$+ \lambda \times \left( \underbrace{\sum_{\left\{\substack{1 \leq a < b \leq N \\ 1 \leq \tau \leq T \\ 1 \leq i \leq k}\right\}} x_{\tau,a}^{(i)} x_{\tau,b}^{(i)} + \sum_{\left\{\substack{1 \leq a \leq N \\ 1 \leq \tau \neq \tau' \leq T \\ 1 \leq i \leq j \leq k}\right\}} x_{\tau',a}^{(i)} x_{\tau,a}^{(j)} + \sum_{\left\{\substack{1 \leq a \leq N \\ 1 \leq \tau \leq T \\ 1 \leq i < j \leq k}\right\}} x_{\tau,a}^{(i)} x_{\tau,a}^{(j)}}_{(*3)} \right).$$

$$\tag{6}$$

---

[1] Note that the unit of time-division can also depend on time $(\tau)$ and vehicle $(i)$, i.e., $\Delta t_\tau^{(i)} \equiv t_{\tau+1}^{(i)} - t_\tau^{(i)}$. Although such a generalization is also important for some applications, we keep the uniform value $\Delta t_\tau^{(i)} = \Delta t$ for the sake of simplicity.

[2] Here $\lceil x \rceil$ $(x \in \mathbb{R})$ is the minimum integer which is greater than or equal to $x$, i.e., $x \leq \lceil x \rceil < x+1$ and $\lceil x \rceil \in \mathbb{Z}$.

The new form of interaction is introduced in the first line: (\*1) and (\*2)

1. The first term (\*1) gives the cost of travel from a city $(b)$ (departed at $\tau$) to a city $(a)$ (the arrival is at $\tau' = \tau + n_{ab}^{(\tau)}$).
2. The second term (\*2) forbids *any early arrivals* at a city $(a)$ (i.e., at $\tau'$ in range $1 \le \tau' - \tau < n_{ab}^{(\tau)}$). This is realized by repulsive interactions set forward from the departure city $(b)$ to the arrival city $(a)$.

The other terms (\*3) are obtained from the basic constraints of penalty terms:

1. Any vehicle $(i)$ in any time-interval $(\tau)$ will not arrive at two different cities $(a)$ and $(b)$ simultaneously:

$$\lambda \times x_{\tau,a}^{(i)} x_{\tau,b}^{(i)} \qquad (\forall a \ne \forall b, \forall i, \forall \tau). \tag{7}$$

2. If a vehicle $(i)$ arrives at a city $(a)$ in a time-interval $(\tau)$, then that vehicle $(i)$ will not arrive at a city $(a)$ in any other time-interval $(\tau')$:

$$\lambda \times x_{\tau,a}^{(i)} x_{\tau',a}^{(i)} \qquad (\forall a, \forall i, \forall \tau \ne \forall \tau'). \tag{8}$$

3. If a vehicle $(i)$ arrives at a city $(a)$ in a time-interval $(\tau)$, then the other vehicles $(j)$ will not arrive at a city $(a)$ in any time-interval $(\tau')$:

$$\lambda \times x_{\tau,a}^{(i)} x_{\tau',a}^{(j)} \qquad (\forall a, \forall i \ne \forall j, \forall \tau, \forall \tau'). \tag{9}$$

At this point, we no longer employ the conventional method of constraint functions (i.e., square of linear-constraint functions, applied in [4]). Instead, we have introduced *an additional parameter* $\mu \, (> 0)$ around the traveling cost $d_{ab}^{(\tau)}$, as well as the standard parameters $\rho \, (> 0)$ and $\lambda \, (> 0)$.

– Overall negative shift of costs (delivered by $\mu$) replaces the role of the negative linear terms generated in the conventional method of constraint functions (See Sect. 6).
– Up to the overall scaling, at least two parameters $(\mu, \rho)$ should be adjusted properly to optimize the performance of each Ising machine. The remaining $\lambda$ is adjusted by total scaling to the maximum value suited for each Ising machine.
– However, as discussed in Sect. 6, the standard value of the parameters $\mu$ and $\rho$ are selected as follows:

$$\mu = d_{max}, \qquad \rho = \frac{d_{max} - d_{min}}{\lambda}. \tag{10}$$

Generally, the initial starting points of the vehicles can be implemented using boundary condition of the binary qubits. For instance, one can choose

$$x_{1,a}^{(i)} = \delta_{a,s_i} \qquad \text{(the starting point of vehicle } (i) \text{ is a city } (s_i)\text{)}. \tag{11}$$

Conversely, to set the final destination, we turn off several binary qubits from which one cannot reach the final destination before exceeding the upper time limit $T$ (where the final destination of vehicle $(i)$ is the city $(e_i)$):

$$x_{\tau,a}^{(i)} = 0 \quad \text{when} \quad \tau + n_{e_i,a}^{(\tau)} > T, \tag{12}$$

Additionally, the forward interactions from the final destination should be replaced as follows:

$$\left( \frac{d_{ab}^{(\tau)} - \mu}{\rho} \times x_{\tau+n_{ab}^{(\tau)},a}^{(i)} \, x_{\tau,b}^{(i)} + \sum_{1 \le \delta\tau \le n_{ab}^{(\tau)} - 1} \lambda \times x_{\tau+\delta\tau,a}^{(i)} \, x_{\tau,b}^{(i)} \right) \Bigg|_{b=e_i}$$

$$\xrightarrow[\text{replace}]{} \left( \sum_{1 \le \delta\tau \le T - \tau} \lambda \times x_{\tau+\delta\tau,a}^{(i)} x_{\tau,e_i}^{(i)} \right), \tag{13}$$

to make sure that any travel after arriving at the final destination is forbidden.

As this TSP/VRP formulation can deal with time-scheduling, we refer it to as *time-scheduled TSP/VRP* or *TS-TSP/VRP*.

## 3    Multiple-Capacitated TSP/VRP

In the last section, we have introduced the concept of time and, as is mentioned in Introduction, time is a kind of capacitated variable whose consumption is accumulated until full of its "capacity". Therefore, one can replace the concept of time by the concept of capacity for "a monotonically increasing/decreasing capacitated variable". This replacement is also useful in some practical applications, especially in a case of saving the number of qubits. However, such a concept of time/capacity only resolves the problem of single-capacitated TSP/VRP.

In this section, we shall introduce the concept of multiple capacities in addition to our time-scheduled TSP/VRP formulation, which is referred to as *time-scheduled multiple-capacitated VRP* or *TS-mCVRP*. Addition of multiple-capacity can be achieved by adding *capacitated variables* $(c_1, c_2, \cdots, c_M)$ in the binary qubits of Eq. (1) as follows:

$$x_{\tau,a|c_1,c_2,\cdots,c_M}^{(i)} \quad \left( 1 \le \tau \le T, \quad 1 \le a \le N, \quad 1 \le i \le k \right), \tag{14}$$

with their capacity bounds given as

$$q_m \le c_m \le Q_m \quad \left( q_m, Q_m \in \mathbb{Z}, \quad 1 \le m \le M \right). \tag{15}$$

In many cases, we use a shorter notation, $c = (c_1, c_2, \cdots, c_M)$, as a vector of $M$-dimensional capacitated variables.[3] The introduction of capacitated variables is

---

[3] We also use the following "array" notation:

$$q \le c \le Q \quad \Leftrightarrow \quad q_m \le c_m \le Q_m \quad (1 \le m \le M). \tag{16}$$

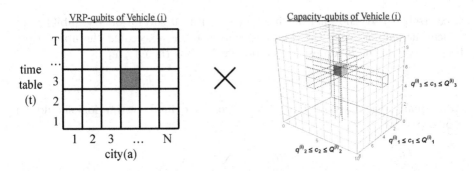

**Fig. 1.** Binary qubits representation as boxes.

understood as the direct product of *capacity-qubits* and the original VRP-qubits (See Fig. 1). The interpretation of the binary values, Eq. (3), should be now incorporated with the status of capacitated variables as follows:

$$x^{(i)}_{\tau,a|c} = 1 \qquad \text{(arriving with the status } c \text{ of capacitated variables)}. \qquad (17)$$

Furthermore, to describe variation of capacitated variables, we introduce *variation matrices* $\left(B^{(\tau)}_{ab|m}\right)_{1 \le a \ne b \le N}$ for each time-interval $(\tau)$:[4]

$$B^{(\tau)}_{ab|m} = \begin{pmatrix} \text{variation of capacitated variable } c_m \text{ for traveling} \\ \text{from a city } (b) \text{ (at time } \tau) \text{ to a city } (a) \end{pmatrix} \in \mathbb{Z}. \qquad (19)$$

We can write the Hamiltonian as follows:

$$\mathcal{H} = \sum_{\left\{ \begin{smallmatrix} 1 \le a \ne b \le N \\ 1 \le \tau \le T-1 \\ 1 \le i \le k \\ q \le c \le Q \end{smallmatrix} \right\}} \begin{pmatrix} \dfrac{d^{(\tau)}_{ab} - \mu}{\rho} \times x^{(i)}_{\tau + n^{(\tau)}_{ab}, a | c + B^{(\tau)}_{ab}} \; x^{(i)}_{\tau, b|c} \\[2ex] + \underbrace{\sum_{q \le c' (\ne c + B^{(\tau)}_{ab}) \le Q} \lambda \times x^{(i)}_{\tau + n^{(\tau)}_{ab}, a | c'} \; x^{(i)}_{\tau, b|c}}_{(*4)} \\[2ex] + \underbrace{\sum_{\substack{1 \le \delta\tau \le n^{(\tau)}_{ab} - 1 \\ q \le c' \le Q}} \lambda \times x^{(i)}_{\tau + \delta\tau, a | c'} \; x^{(i)}_{\tau, b|c}}_{(*5)} \end{pmatrix} \qquad (20)$$

$$+ \lambda \times \begin{pmatrix} \underbrace{\sum_{\left\{ \begin{smallmatrix} 1 \le a \le N \\ 1 \le \tau \le T \\ 1 \le i \le k \\ q \le c' < c \le Q \end{smallmatrix} \right\}} x^{(i)}_{\tau, a|c'} \; x^{(i)}_{\tau, a|c}}_{(*6)} + \sum_{\left\{ \begin{smallmatrix} 1 \le a < b \le N \\ 1 \le \tau \le T \\ 1 \le i \le k \\ q \le c', c \le Q \end{smallmatrix} \right\}} x^{(i)}_{\tau, a|c'} \; x^{(i)}_{\tau, b|c} \\[2ex] + \sum_{\left\{ \begin{smallmatrix} 1 \le a \le N \\ 1 \le \tau \ne \tau' \le T \\ 1 \le i \le j \le k \\ q \le c', c \le Q \end{smallmatrix} \right\}} x^{(i)}_{\tau', a|c'} \; x^{(j)}_{\tau, a|c} + \sum_{\left\{ \begin{smallmatrix} 1 \le a \le N \\ 1 \le \tau \le T \\ 1 \le i < j \le k \\ q \le c', c \le Q \end{smallmatrix} \right\}} x^{(i)}_{\tau, a|c'} \; x^{(j)}_{\tau, a|c} \end{pmatrix}.$$

---

[4] It is also convenient to use a vector notation of the collection of capacity-variation matrices:

$$B^{(\tau)}_{ab} \equiv \left(B^{(\tau)}_{ab|1}, B^{(\tau)}_{ab|2}, \cdots, B^{(\tau)}_{ab|M}\right). \qquad (18)$$

As is in Eq. (6), the first line includes the interaction associated with consumption of time and variation of capacitated variables. In particular,

- The second term (∗4) in the first line ensures that transmission of capacitated variables in each travel satisfies the expected variation relation,

$$c \quad \mapsto \quad c' = c + B_{ab}^{(\tau)}, \tag{21}$$

- The third term (∗5) forbids early arrivals at a city $(a)$ from a city $(b)$ with any changes of capacitated variables, $c \rightarrow c'$.

The second part represents the basic constraints of penalty terms:

1. The newly introduced term is the first term (∗6) and it represents that a vehicle $(i)$ arriving at a city $(a)$ in a time-interval $(\tau)$ cannot be assigned more than one capacitated-variable status, i.e., $c' \neq c$.

$$\lambda \times x_{\tau,a|c'}^{(i)} \, x_{\tau,a|c}^{(i)} \qquad \left( \forall a, \forall i, \forall \tau, \forall c' \neq \forall c \right). \tag{22}$$

2. The other terms are essentially the same as in the previous section.

As is in the previous section as well as Sect. 6, the parameters $(\mu, \rho, \lambda)$ should be chosen as the same standard values as stated in Eq. (10).

A new feature obtained by our capacity formulation is that the variation matrices $B_{ab|m}^{(\tau)}$ can take any integer number. Therefore, the capacitated variables can increase and/or decrease (i.e., pickup and/or delivery), strictly satisfying the capacity bounds.

## 4   State of Vehicles

In addition to the simple concept of capacity, the capacitated variables can also be interpreted as *the states of vehicles*. Through the use of "state", we can introduce various traveling rules depending on each state of vehicles. For the sake of simplicity, we shall demonstrate *two-state TSP/VRP* here.

As a simple example of a two-state TSP/VRP, we shall consider the arrival-state and departure-state of the vehicles. In terms of qubits, the states of vehicle are now denoted by the hat "^" above the binary variables:

$$\widehat{x}_{\tau,a}^{(i)} : \quad \text{arrival-qubits,} \qquad x_{\tau,a}^{(i)} : \quad \text{departure-qubits.} \tag{23}$$

A new ingredient of introducing two states is that we can now describe traveling/staying phases of vehicle:

$$(1) \quad x \rightarrow \widehat{x} : \quad \text{traveling phase,} \qquad (2) \quad \widehat{x} \rightarrow x : \quad \text{staying phase,} \tag{24}$$

and other transitions, say $x \rightarrow x$ and $\widehat{x} \rightarrow \widehat{x}$, should be forbidden. As an additional phase of motion is introduced, we can further add a new time-duration matrix $\widehat{n}_a^{(\tau)}$ to describe *how long the vehicle stays at the city $(a)$*:

$$n_a^{(\tau)} = \left\lceil \frac{(\text{time spent for staying at a city } (a) \text{ before departure})}{\Delta t} \right\rceil \geq 1. \tag{25}$$

Therefore, various traveling rules are additionally introduced as follows:[5]

$$
\begin{aligned}
\mathcal{H} = & \sum_{\left\{\substack{1\leq a\neq b\leq N \\ 1\leq\tau\leq T-1 \\ 1\leq i\leq k}\right\}} \left( \underbrace{\frac{d_{ab}^{(\tau)}-\mu}{\rho} \times \widehat{x}_{\tau+n_{ab}^{(\tau)},a}^{(i)}\, x_{\tau,b}^{(i)} + \sum_{1\leq\delta\tau\leq n_{ab}^{(\tau)}-1} \lambda \times \widehat{x}_{\tau+\delta\tau,a}^{(i)}\, x_{\tau,b}^{(i)}}_{\text{``traveling''}} \right. \\
& \left. + \sum_{1\leq\delta\tau\leq n_{ab}^{(\tau)}} \lambda \times x_{\tau+\delta\tau,a}^{(i)}\, x_{\tau,b}^{(i)} \right) \\
+ & \sum_{\left\{\substack{1\leq a,b\leq N \\ 1\leq\tau\leq T-1 \\ 1\leq i\leq k}\right\}} \left( \underbrace{\delta_{ab}\left(\frac{0-\mu}{\rho} \times x_{\tau+n_{a}^{(\tau)},a}\widehat{x}_{\tau,a}^{(i)} + \sum_{1\leq\delta\tau\leq n_{a}^{(\tau)}-1} \lambda \times x_{\tau+\delta\tau,a}^{(i)}\widehat{x}_{\tau,a}^{(i)}\right)}_{\text{``staying''}} \right. \\
& \left. + (1-\delta_{ab}) \sum_{1\leq\delta\tau\leq n_{b}^{(\tau)}} \lambda \times \widehat{x}_{\tau+\delta\tau,a}^{(i)}\widehat{x}_{\tau,b}^{(i)} \right) \\
+ & \lambda \times \left( \sum_{\left\{\substack{1\leq a\leq N \\ 1\leq\tau\leq T \\ 1\leq i\leq k}\right\}} \widehat{x}_{\tau,a}^{(i)}\, x_{\tau,a}^{(i)} + \sum_{\left\{\substack{1\leq a<b\leq N \\ 1\leq\tau\leq T \\ 1\leq i\leq k}\right\}} \left(x_{\tau,a}^{(i)}\, x_{\tau,b}^{(i)} + \widehat{x}_{\tau,a}^{(i)}\, x_{\tau,b}^{(i)}\right) \right. \\
& + \sum_{\left\{\substack{1\leq a\leq N \\ 1\leq\tau\neq\tau'\leq T \\ 1\leq i\leq j\leq k}\right\}} \left(x_{\tau',a}^{(i)}\, x_{\tau,a}^{(j)} + \widehat{x}_{\tau',a}^{(i)}\, x_{\tau,a}^{(j)}\right) \\
& + \sum_{\left\{\substack{1\leq a\leq N \\ 1\leq\tau\leq T \\ 1\leq i<j\leq k}\right\}} \left(x_{\tau,a}^{(i)}\, x_{\tau,a}^{(j)} + \widehat{x}_{\tau,a}^{(i)}\, x_{\tau,a}^{(j)}\right) \\
& + \left( \text{``hermitian conjugate''} \right) \Bigg).
\end{aligned}
\tag{26}
$$

Construction of the Hamiltonian is essentially the same as that of capacity (particularly the last line, which was obtained by re-interpreting capacitated variables $c$ of Sect. 3 as states). A distinguishing point is that we can introduce different cost matrices $d_{ab}^{(\tau)}$ and duration matrices $n_{ab}^{(\tau)}$ depending on the state (i.e., phase) of each vehicle.

This system (including "states" of vehicle) is referred to as *time-scheduled state-vehicle routing problem* or *TS-SVRP*. As a trivial extension, one can also consider multiple-state models and also the full system of *time-scheduled multiple-capacitated state-vehicle routing problem* or *TS-mCSVRP*.

## 5    Practical Constraints with Time, State and Capacity

In practical applications for mobility service, there are various kinds of constraints that should be *simultaneously* implemented. The following list describes examples of such constraints, accepted by our TS-mCSVRP formulation:

(1) **setting:**  Consider an optimization problem for a delivery service. Cost of delivery is the total delivery time, and there are 50 customers, which are shared by five vehicles.

---

[5] "Hermitian conjugate" of the variables is defined as: $\widehat{x_{\tau,a}^{(i)}} = \widehat{x}_{\tau,a}^{(i)}$ and $\widehat{\widehat{x}_{\tau,a}^{(i)}} = x_{\tau,a}^{(i)}$.

- Five vehicles comprises three middle-size cars and two trucks.
- Delivery is served in eight hours (e.g., from 9:00 to 17:00).
- Delivery schedule is set by a twenty-minute unit.
- The cost of optimization would be the total consumption of delivery time.

(2) **time-variation:**    Traffic conditions can change depending on the delivery time. Therefore, the cost $d_{ab}^{(\tau)}$ depends on time $\tau$. Similarly, the cost of inbound and outbound routes is different in general, i.e., $d_{ab}^{(\tau)} \neq d_{ba}^{(\tau)}$.

(3) **priority for delivery:**    Some customers may request a priority for delivery, which can be achieved by adding some extra weight factor to the cost function, $d_{ab}^{(\tau)} \rightarrow \varphi * d_{ab}^{(\tau)}$ $(0 < \varphi < 1)$.

(4) **time-window and type-window:**    There are three kinds of constraints associated with *window*. These constraints can be realized by turning off the associated binary qubits (i.e., $x_{\tau,a|c}^{(i)} = 0$) by hand:

- Each customer has a request of *delivery time*, scheduled by twenty minutes. The delivery-time request can be multiple (i.e., disconnected) time-window for each customer.
- Some of the vehicles cannot deliver to some customers; e.g. the roads are too narrow for trucks and some of the packages are too large or heavy for small cars to deliver. This requires *type-window* for delivery service.
- Each vehicle should serve within their own working hours. In particular, some of the drivers work in a short time. This requires *time-window for vehicles*.

Notably, only the formulation of time-window has a discrepancy with our capacity and state description as it induces overtime traveling sometimes (See Sect. 6). Except for time-window, one can further implement the following two constraints:

(5) **capacity-constraints:**    Capacity constraints are realized using capacity-qubits:

- Each vehicle has its own volume and weight limitations for their capacity.
- For some vehicles that are used for both pickup and delivery, they need to be scheduled without over-capacity for volume and weight in delivering their service.

(6) **scheduling as state:**    Some detail about the schedule can be described using *state-description*.

- After arriving at each customer's premise, drivers should spend twenty minutes on their customer services.
- Drivers can take a one-hour rest for lunch. This can be formulated using the time-dependence of duration matrices $n_{ab}^{(\tau)}$ or $n_a^{(\tau)}$.

# 6   Validity of the Formulation with D-Wave 2000Q

Thus far, we have discussed the proposed new QUBO formulation of TSP/VRP. Conversely, it is also important to discuss the validity of the proposed formulation using a quantum annealing machine, D-Wave 2000Q.

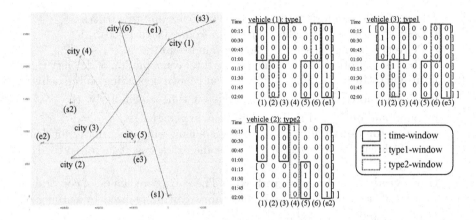

**Fig. 2.** Graphical view of the solution for a VRP-instance: the lowest energy state found with `num_reads` = 10,000 (i.e., time-to-solution, tts $\simeq 20\,\mu\text{sec} \times 10,000 = 0.2\,\text{sec}$) on the D-Wave QPU, `DWave2000Q_2_1`. Minor embedding is processed by the `find_embedding` utility provided by D-Wave System Inc., where `minimize_energy` is used for the broken chain treatment. Among the 10,000 samples, 80% of the solutions are feasible solutions, which do not receive any penalty contributions.

Figure 2 shows a primitive instance of a delivery service (TS-VRP with windows) for six customers with three vehicles in two hours. The unit of time-scheduling is chosen to be fifteen minutes. Interestingly, a travel from the city (4) to the city (5) automatically chooses overtime traveling, because time-window occasionally prevents the shortest travel.

Note that the parameters $(\rho, \mu, \lambda)$ are chosen as the standard value, Eq. (10), and therefore further optimization of the parameters would improve the performance. With taking into account this result, this instance shows that quantum annealing of our formulation works properly as far as small-size QUBO systems ($\simeq 83$ logical qubits for this instance) are considered.

## 6.1   Negatively-Shifted Energy Method and Choice of Baselines

Further clarification is needed about our proposed method for the basic constraints (discussed around Eq. (10)). As mentioned earlier, we did not apply the conventional method of "square of linear-constraint functions," as in [4]. Instead, we apply *a negatively-shifted energy method*; therefore we will discuss how our proposed method can replace the conventional method.

We consider a typical Hamiltonian, which can be generally expressed as follows:

$$\mathcal{H} = \sum_{A,B=1}^{N} \underbrace{(C_{AB} - \mu)}_{\equiv \epsilon_{AB}} \times x_A\, x_B - \xi \sum_{A=1}^{N} x_A + \sum_{\langle A,B \rangle \in E_{\text{frb.}}} \lambda \times x_A\, x_B, \qquad (27)$$

where $C_{AB}$ $(> 0)$ is the general cost matrix and $E_{\text{fbd}}$ is a set of forbidden pairs (i.e., edges) of configurations. We usually choose $\xi = 0$.

In the conventional method (See [4]), the parameter $\xi$ inevitably exists and is strictly correlated with $\lambda$ $(> 0)$ as follows:

$$\xi = \lambda \quad \text{(TSP)}, \qquad \xi \geq \frac{3}{2}\lambda \quad \text{(VRP)}, \tag{28}$$

and $\mu = 0$. If one forgets about the correlation and chooses $\xi = 0$ and $\lambda > 0$ (and $\mu = 0$ in the conventional method), the ground state becomes trivial, $x_A = 0$ $(A = 1, 2, \cdots, N)$. Therefore, the basic role of $\xi$-term is to enhance spontaneous popping-up of qubits: $x_A \to 1$ $(A = 1, 2, \cdots, N)$.

However, such effect can also be generated using *the overall negative shift of cost energy* obtained from $\mu$. This is possible since optimization only cares about the relative values of cost. Hence we can adjust $\mu$ such that

$$\epsilon_{AB} = C_{AB} - \mu < 0, \tag{29}$$

for all the pairs $\langle A, B \rangle$ focused.[6] In the proposed method, we can set $\xi = 0$ or utilize it for other purposes.

In a sequence of configurations generated using the negative energy, Eq. (29), there are forbidden configurations, which receive penalty caused by $\lambda$. As these configurations are forbidden, we impose that such configurations receive relatively positive energy. In our proposed VRP, this condition is given by

$$\epsilon_{AB} + \lambda > 0, \tag{30}$$

for all pairs $\langle A, B \rangle$. We put a baseline (like a coastline) as energy $= 0$ to separate feasible configurations (inside sea) from forbidden configurations (on the continent). This is referred to as the baseline condition.

From these two conditions, we choose the maximum range of cost energy,

$$-\lambda = \epsilon_{\min} \leq \epsilon_{AB} \leq \epsilon_{\max} = 0, \tag{31}$$

as the standard values. The larger range of cost causes larger resolution in quantum annealing machines. The solution of these conditions is given by Eq. (10).

It is also possible to strategically choose higher baseline in Eq. (30) and/or to select some focused energy range of Eq. (29) as

$$\epsilon_{AB} + \lambda + \delta > 0 \quad \Rightarrow \quad -\lambda - \delta = \epsilon_{\min} \leq \epsilon_{AB} \leq \epsilon_{\max}^{(\text{focused})} = 0, \tag{32}$$

which may improve the performance. Despite increase in forbidden configurations, it can further improve the range of energy input. In contrast, too large $\delta (> 0)$ reduces performance, because the probability of ground states evaporates into the forbidden configurations, which possess much larger entropy.[7]

---

[6] This means that $\epsilon_{AB} \leq C_{\max}^{(\text{focused})} - \mu = 0 \leq C_{\max} - \mu$ for all the focused pairs $\langle A, B \rangle$. Therefore, the parameter $\mu$ plays a role of "cut-off scale" of cost energy.

[7] This phenomenon also occurs in the conventional method: if the parameter $\xi$ becomes too large, it induces a large number of forbidden configurations. Therefore, too large $\xi$ also reduces the performance of quantum annealing.

# 7    Conclusion and Discussion

In this paper, we proposed a new QUBO formulation of CVRP with time, state, and capacity. Introduction of the strict concept of time allowed us to formulate various constraints associated with time, while the introduction of capacity-qubits allowed us to formulate pickup and delivery during each travel of the vehicles. Introduction of state allowed us to describe various traveling rules depending on the state of the vehicles. We evaluated the proposed QUBO formulation using a quantum annealing machine, D-Wave 2000Q and the results show that our formulation properly works in small-size QUBO systems (less than 90 logical qubits $\simeq 6 \sim 7$ customers), which can be directly embedded in the current D-Wave machines.

For real-world applications of this formulation, on the other hand, at least more than 2000 logical qubits ($\simeq$ more than 30 customers for 20-minute scheduling of a half day) are required. In this sense, it is also interesting to evaluate our proposed TS-mCSVRP on digital Ising machines. It is also important to develop an efficient quantum/classical hybrid algorithm for our TS-mCSVRP, which should hasten the practical usage of our formulation. This point shall be reported in the future communication.

# References

1. Dantzig, G.B., Ramser, J.H.: The truck dispatching problem. Manag. Sci. **6**(1), 80–91 (1959). https://doi.org/10.1287/mnsc.6.1.80
2. Kadowaki, T., Nishimori, H.: Quantum annealing in the transverse Ising model. Phys. Rev. E **58**, 5355–5363 (1998). https://doi.org/10.1103/PhysRevE.58.5355
3. Johnson, M.W., et al.: Quantum annealing with manufactured spins. Nature **473**, 194–198 (2011)
4. Lucas, A.: Ising formulations of many NP problems. Front. Phys. **2**, 1–5 (2014). https://doi.org/10.3389/fphy.2014.00005
5. Hopfield, J.J., Tank, D.W.: Neural computation of decisions in optimization problems. Biol. Cybern. **52**, 141–152 (1985). https://doi.org/10.1007/BF00339943
6. Hopfield, J.J., Tank, D.W.: Computing with neural circuits: a model. Sci. New Ser. **233**(4764), 625–633 (1986). https://doi.org/10.1126/science.3755256
7. Mèzard, M., Parisi, G.: Replicas and optimization. J. Phys. Lett. **46**, 771–778 (1985). https://doi.org/10.1051/jphyslet:019850046017077100
8. Itoh, T., Ohta, M., Yamazaki, Y., Tanaka, S.: Quantum annealing for combinatorial optimization problems with multiple constraints. In: A poster in Adiabatic Quantum Computing Conference, AQC-17, 26 June 2017, Tokyo, Japan (2017)
9. Feld, S., Gabor, T.: As a jointwork with Volkswagen: A talk in Qubits Europe 2018 D-Wave Users Conference, 12 April 2018, Munich, Germany. https://www.dwavesys.com/sites/default/files/lmu-merged-published.pdf

# Boosting Quantum Annealing Performance Using Evolution Strategies for Annealing Offsets Tuning

Sheir Yarkoni[1,2]($\boxtimes$), Hao Wang[2], Aske Plaat[2], and Thomas Bäck[2]

[1] D-Wave Systems Inc., Burnaby, Canada
syarkoni@dwavesys.com
[2] LIACS, Leiden University, Leiden, Netherlands

**Abstract.** In this paper we introduce a novel algorithm to iteratively tune annealing offsets for qubits in a D-Wave 2000Q quantum processing unit (QPU). Using a (1+1)-CMA-ES algorithm, we are able to improve the performance of the QPU by up to a factor of 12.4 in probability of obtaining ground states for small problems, and obtain previously inaccessible (i.e., better) solutions for larger problems. We also make efficient use of QPU samples as a resource, using 100 times less resources than existing tuning methods. The success of this approach demonstrates how quantum computing can benefit from classical algorithms, and opens the door to new hybrid methods of computing.

**Keywords:** Quantum computing · Quantum annealing ·
Optimization · Hybrid algorithms

## 1 Introduction

Commercial quantum processing units (QPUs) such as those produced by D-Wave Systems have been the subject of many characterization tests in a variety of optimization, sampling, and quantum simulation applications [1–6]. However, despite the increasing body of work that showcases the various uses of such QPUs, application-relevant software that uses the QPUs intelligently has been slow to develop. This can be most easily attributed to the fact that these QPUs have a multitude of parameters, with each contributing (possibly in a co-dependent manner) to the performance of the processor. In practice it is intractable to tune all parameters, and often a subset of parameters are chosen to be explored in detail. In this paper we introduce the use of a Covariance Matrix Adaptation Evolution Strategy (CMA-ES) to heuristically tune one set of parameters, the so-called annealing offsets, on-the-fly, in an application-agnostic manner. We test the viability and performance of our tuning method using randomly generated instances of the Maximum Independent Set (MIS) problem, a known NP-hard optimization problem that has been tested on a D-Wave QPU before [7]. The definition of the MIS problem is as follows: given a graph $G$ composed of vertices

© Springer Nature Switzerland AG 2019
S. Feld and C. Linnhoff-Popien (Eds.): QTOP 2019, LNCS 11413, pp. 157–168, 2019.
https://doi.org/10.1007/978-3-030-14082-3_14

$V$ and edges $E$ between them, find the largest subset of nodes $V' \subseteq V$ such that no two nodes in $V'$ contain an edge in $E$. The tuning method introduced in this paper does not exploit the nature of the MIS problem nor makes any assumptions regarding the structure of the associated graph, and can be used for problem sets other than MIS instances.

The QPUs available from D-Wave Systems implement a quantum annealing algorithm which samples from a user-specified Ising Hamiltonian (in a $\{-1, +1\}$ basis) or equivalently a quadratic unconstrained binary optimization (QUBO) problem (in a $\{0, 1\}$ basis) [8]. Minimizing a QUBO or Ising Hamiltonian is known to be an NP-hard problem [9], so many interesting and difficult computational problems can be formulated as Ising models and QUBOs [10]. A QUBO problem and its associated objective function can be represented as the following:

$$\text{Obj}(x) = x^T \cdot Q \cdot x, \tag{1}$$

where $Q$ is a matrix $\in \mathbb{R}^{N \times N}$, and $x$ is a binary string $\in \{0, 1\}^N$.

The rest of the paper is organized as follows. In the next section we introduce the previous works regarding annealing offsets in D-Wave QPUs. Section 3 describes the motivation and use of annealing offsets as implemented by D-Wave. In Sect. 5 we explain in detail the algorithm implemented in this paper: how the annealing offsets were evolved and then used to solve randomly generated MIS problems. The results obtained using this algorithm are shown in Sect. 6, and are followed by conclusions in Sect. 7.

## 2   Previous Works

It has been shown experimentally in [11] that the probability of staying in the ground state of the quantum system can be improved by applying annealing offsets to small qubit systems. Specifically, the authors use a first-order perturbative expansion of the quantum annealing Hamiltonian to connect ground state degeneracy and single-spin floppiness to change the annealing offset for particular qubits. Small instances were chosen (24 qubits) such that exact diagonalization of the quantum Hamiltonian was possible, which is necessary for the iterative algorithm presented in [11]. For these instances, which were chosen specifically to be difficult for quantum annealing, median success probability of the QPU was improved from 62% to 85% using an iterative algorithm. In a more recent white paper produced by D-Wave Systems [12], it was shown that for specific input sets it is possible to boost performance of the QPU by up to a factor of 1000. These results were obtained by using a grid-search technique, requiring 2.5 million samples to be generated using the QPU for a single input problem. The model used to adjust the annealing offsets in this paper was similar to the model used in [11]. This technique was also applied to 2-SAT problems with 12 Boolean variables in [13].

It has been shown (using quantum Monte Carlo simulations) that annealing offsets can mitigate first-order phase transitions, exponentially enhancing performance when compared to simulated thermal annealing [14]. However, this work used simple problems that could be solved exactly. Understanding how this asymptotic behavior affects NP-hard problems in general is an open question. Previous work regarding solving maximum independent set problems using a D-Wave 2000Q QPU shows that the QPU performance suffers greatly as problem sizes increase, even as problems become easier to solve classically [7]. The new annealing offset features in the D-Wave 2000Q processor are designed to mitigate the likely sources of this performance degradation seen in [7]. Tuning these parameters using a CMA-ES routine is the subject of this work.

## 3   Quantum Annealing Offset Parameters

In uniform quantum annealing as implemented in D-Wave QPUs [8, 15], all qubits begin their evolution at the same point in time. Consider an annealing procedure defined as follows:

$$H(s) = A(s) \sum_i \sigma_i^x + B(s) \left[ \sum_i h_i \sigma_i^z + \sum_{ij} J_{ij}\, \sigma_i^z \otimes \sigma_j^z \right],$$

where $A(s)$ is the initial Hamiltonian driver function (transverse field), $B(s)$ is the target Hamiltonian driver function (terminal energy scale), $h$ and $J$ define the target Hamiltonian, $s$ is a normalized time parameter (real time $t$ divided by total anneal time $\tau$), and $\sigma_i^x$ and $\sigma_i^z$ are the $x$ and $z$ Pauli spin matrices operating on the $i$th qubit respectively. By default, all qubits will begin their annealing procedure at time $s = 0$, and terminate at $s = 1$. However, a feature in the D-Wave 2000Q processor allows users to set an advance or delay for qubits in physical time, and in an independent manner. Meaning, each qubit can have its evolution path advanced/delayed by some user specified[1] $\Delta s$. Thus the functions $A(s)$ and $B(s)$ are now an ensemble of functions, $A(s + \Delta s_i)$ and $B(s + \Delta s_i)$, ranging from $s = 0 + \Delta s_i$ to $s = 1 + \Delta s_i$, and can differ from qubit to qubit.

The benefit of adding this additional control feature is motivated by the physics governing quantum annealing. As demonstrated in [4], the distribution of answers produced by the QPU are more than a result of the finite temperature Boltzmann distribution, but are also an artifact of ergodicity breaking during the annealing procedure, and are a result of the annealing path as defined by the $A(s)$ and $B(s)$ functions in Eq. 3. Allowing a programmable method to delay and advance the anneal schedule of individual qubits allows some mitigation of this effect, sometimes called "freeze-out". It has even been shown that, for careful construction of these offsets on a per-qubit basis, it may be possible to avoid forbidden crossings entirely [14]. However, this requires *a priori* knowledge of the energy landscape, and is thus computationally impractical for most problems.

---

[1] Each qubit in the QPU has a different range in $\Delta s$ that can be set independently. $\Delta s < 0$ is an advance in time and $\Delta s > 0$ is a delay. A typical range for $\Delta s$ is $\pm 0.15$. The total annealing time for all qubits is still $|s| = 1$ (or $\tau$ in units of time).

## 4    Covariance Matrix Adaptation Evolution Strategy (CMA-ES)

The Covariance Matrix Adaptation Evolution Strategy (CMA-ES) [16] is a state-of-the-art stochastic optimization algorithm for the continuous black-box optimization problem. To tune the annealing offsets, we adopt the so-called $(1 + 1)$-CMA-ES variant [17], which generates only one candidate search point in each iteration. The $(1 + 1)$-CMA-ES algorithm exhibits both fast convergence and global search ability. The choice of the optimization algorithm is made based on the following considerations: firstly, QPU time is considered an expensive resource, and we wish to spend as little QPU time for the tuning as possible. Secondly, the QPU is a serial device, meaning that problems (a set of annealing offset parameters in our case) can only be tested sequentially. Therefore, the search strategies that generate multiple candidate points are not preferred as it is not possible to parallelize those points in our case.

For the experiments reported in this paper, we run the $(1 + 1)$-CMA-ES with its default parameter settings, since it is well known that such a setting shows quite robust behaviors across many benchmark functions [18]. Only the method for generating the initial candidate solution is varied in the experiment: we investigate the effects of initializing the annealing offsets either uniformly within the corresponding upper/lower bounds, or with a value of zero.

## 5    Tuning QA Parameters with (1+1)-CMA-ES

In this paper, it is proposed to tune the annealing offsets of a D-Wave QPU with the $(1+1)$-CMA-ES algorithm [17], aiming at improving the performance of the QPU on specific problem classes, for instance the MIS problem. MIS problem instances are constructed using the same method as in [7], using random graphs with edge probability $p = 0.2$ (the empirically determined difficult point for the QPU). The minor-embedding techniques applied to solve these problems directly on the QPU topology are also as in [7]. To solve these problems, the annealing offsets of qubits should be set up properly and thus are tuned using $(1+1)$-CMA-ES. We considered the qubits within each chain to be a single logical qubit, as with the purpose of embedding. Therefore, in calculating the offsets, we consider the *collective* minimum and maximum offsets that can be used on the chain as a whole. Explicitly, we find the highest minimum and lowest maximum shared between all qubits in every chain, and use those as the boundary for each logical qubit (chain). Every qubit within every chain is therefore advanced/delayed by the same amount.

In the tuning experiment, two initialization methods of annealing offsets are compared: *uniform* and *none*. The former is a relatively standard approach: given a feasible range for each offset, $[l_i, u_i]$ for qubits in chain $c_i$, a random number is sampled uniformly within this range, which is then used as the initial search point for all qubits in the chain. This method allows for a fair exploration of the entire search space. It is especially important in such a quadratic model, where the influences of each annealing offset on the others are not obvious and unpredictable in the worst case. The second method, where all offsets are set to zero initially, is also tested because it represents a "not bad" initial position, and is also the starting point for the algorithm presented in [11]. This is the default setting when using D-Wave QPUs. We consider this point to be a local optimum for annealing offsets and attempt to improve upon this optimum using the CMA-ES procedure.

The fitness function (objective function) of $(1+1)$-CMA-ES is calculated as the mean energy of the solutions returned by the QPU. This was observed to be the most stable fitness function (as opposed to the 25% percentile, or the minimum energy). Due to the stochastic nature of the samples returned by the QPU, it is important to use a stable metric to evaluate the fitness function in every iteration. According to some preliminary tests, other metrics are too noisy to enable the fast convergence of the $(1+1)$-CMA-ES algorithm. Examples of tuning runs are presented in Appendix A.

---

**Algorithm 1.** Tune annealing offsets using $(1+1)$-CMA-ES

---

1: **procedure** TUNE-OFFSET($l, u, B, StepCost$)
2:     Initialize: $\sigma \leftarrow \max\{u - l\}/4$, $\mathbf{C} = \mathbf{I}$, $c \leftarrow 0$
3:     **if** InitialOffsets $= 0$ **then**
4:         $\mathbf{x} \leftarrow \mathbf{0}$                                                                    ▷ offset: zero initialization
5:     **else**
6:         $\mathbf{x} \leftarrow \mathcal{U}(l_i, u_i)$                                                       ▷ offset: uniform initialization
7:     $f(\mathbf{x}) \leftarrow$ CALL-QPU($\mathbf{x}, StepCost$)
8:     $\mathbf{A}\mathbf{A}^\top \leftarrow \mathbf{C}$                                                        ▷ Cholesky decomposition
9:     **while** $c < B$ **do**
10:         $\mathbf{z} \leftarrow \mathcal{N}(\mathbf{0}, \mathbf{I})$                                        ▷ standard normal distribution
11:         $\mathbf{x}' \leftarrow \mathbf{x} + \sigma \mathbf{A}\mathbf{z}$
12:         $f(\mathbf{x}') \leftarrow$ CALL-QPU($\mathbf{x}', StepCost$)
13:         $\sigma \leftarrow$ UPDATE-STEP-SIZE($\sigma$)
14:         **if** $f(\mathbf{x}') < f(\mathbf{x})$ **then**
15:             $\mathbf{x} \leftarrow \mathbf{x}'$
16:             $\mathbf{A} \leftarrow$ UPDATE-CHOLESKY($\mathbf{A}, \mathbf{z}$)
17:         $c \leftarrow c + StepCost$
18:     **return x**

---

Pseudocode outlining the algorithm used to tune the offsets is shown in Algorithm 1 and the appropriate terms (along with the values used in our experiments, when applicable) are defined in Table 1. Essentially, the proposed algorithm optimizes the annealing offsets using the so-called mutation operation (Line 10 and 11), where the current annealing offset $\mathbf{x}$ is perturbed by a Gaussian random vector $\sigma \mathbf{A}\mathbf{z}$ (which is transformed and rescaled from the standard normal vector $\mathbf{z}$, Line 10). The resulting mutation $\mathbf{x}'$ is evaluated in the QPU (Line 12) and it is passed onto the next iteration if its objective value $f(\mathbf{x}')$ is better than $f(\mathbf{x})$ (Line 14 and 15). In addition, two procedures, UPDATE-STEP-SIZE and UPDATE-CHOLESKY are adopted to control the step-size $\sigma$ and the matrix $\mathbf{A}$. The details of those two procedures are presented in [17]. After the depletion of the total budget, we use the final annealing offsets produced by CMA-ES to solve the problem instances: 10,000 samples are assigned to the QPU and the final annealing offsets **x** from Algorithm 1 are used. Given a budget of 20,000 samples per MIS instance, this means that 50% of the budget (per MIS instance) is allocated for offset tuning while the remaining 50% are used for problem solving. The goal is to allocate a sufficient amount of resources for calibration, and then solve the MIS problems with the remaining QPU time.

**Table 1.** Explanation of variables and procedure used in Algorithm 1.

| | |
|---|---|
| $B$ | Total budget (amount of resources) for tuning QPU annealing offsets (in units of total number of samples drawn from the QPU; In total, 10,000 samples are used in our experiment per instance.) |
| InitialOffsets | The initial value for each offset of qubits in the problem; we test either all set to 0 or uniformly between their min/max range |
| StepCost | The cost of each step of the fitness evaluation of the offsets (in number of samples from the QPU; we used 100 samples per call) |
| **x** | The current annealing offsets |
| $f(\mathbf{x})$ | Fitness value of the current offsets, measured in units of mean energy of the samples returned by the QPU |
| $c$ | Counter for the budget (measured in number of samples from the QPU) |
| CALL-QPU | The objective function that calls the QPU, takes **x** and StepCost as arguments |
| $\sigma$ | The step-size that scales the mutation of offsets |
| **C** | The matrix of covariances between the annealing offset values |
| **A** | The Cholesky decomposition of **C** |
| UPDATE-STEP-SIZE | The procedure to control the step-size $\sigma$. Please see [17] for the detail |
| UPDATE-CHOLESKY | The procedure to adapt the Cholesky decomposition **A** of the covariance matrix **C**. Please see [17] for details |
| $\mathcal{U}(a, b)$ | Uniform random distribution in $[a, b]$ |
| $\mathcal{N}(\mathbf{0}, \mathbf{I})$ | Standard multivariate normal distribution |

# 6 Experimental Results

Similar to previous work [7], 50 random graphs with 20–60 variables were generated using edge probability $p = 0.2$, and the MIS problem was solved for each. The QPU annealing offsets were tuned using 10,000 samples per instance as described in Sect. 5. Each tuned annealing offset configuration was then used to collect 10,000 samples from the QPU for each MIS instance, as well as the "no offsets" configuration for comparison. Figure 1 (left) shows the mean probability of success for configurations that found the presumed optimum. We show the two configurations of tuning (without initial offsets and uniform offsets) relative to the naive performance of no tuning. It is possible for certain configurations to not find the global optimum, even after tuning, as shown in Fig. 1 (right). As expected, using the QPU with no annealing offsets leads to the highest amount of unsolved instances at the largest problem sizes. This means that, on average, the tuned annealing offsets find optima (larger independent sets) that *cannot be obtained without tuning.* Surprisingly, however, there are problem instances that remain unsolved even for smaller problem sizes with tuning. Additionally, the initial point for the CMA-ES routine (i.e., uniform vs. null initial offsets) affects the number of unsolved problems at smaller sizes. This is likely due to the uniform offsets forcing the QPU into a bad initial configuration without sufficient resources for the CMA-ES routine to correct this.

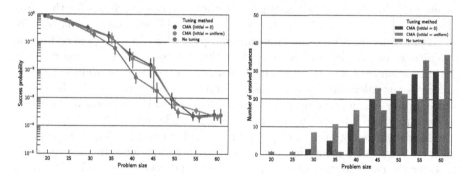

**Fig. 1. Left:** Mean success probability (of instances where tuned solvers found the optimum) for increasing problem sizes. Higher success probabilities are better, indicating improved performance in finding optima. **Right:** Bar chart showing how many problems remained unsolved by the solvers, before/after tuning with different configurations. As problem sizes increase, the no tuning version of the QPU solves relatively fewer instances.

The null initial offsets, which can be viewed as a stable local minimum for smaller problems, typically solve more instances than the uniform initial offsets configuration, although both are outperformed by no offsets (Fig. 1, right). This is consistent with previous observations [7] where the QPU was able to outperform even simulated thermal annealing at small problem sizes. In the cases where both tuning configurations found the optimum, Fig. 1 shows that the probability of obtaining ground states is similar in both configurations, and both versions outperformed the QPU without tuning for all problem sizes. In Fig. 2 we show the ratio of the mean success probability of instances where the ground state was obtained between the tuned and untuned offsets. This measures the improvement in probability of obtaining the ground state for a particular MIS instance. The improvement obtained by using the two configurations is qualitatively similar, and peaks at 12.4 times improvement at problem size 45 for the null initial offsets, and a factor of 8.2 improvement at problem size 45 for the uniform initial offsets. The improvement gained by tuning the annealing offsets steadily increases with problem size, until reaching its peak, after which the gains mostly disappear (although we are still able to solve a higher number of problems after tuning). This behavior indicates that at small problem sizes there is little to be gained from tuning, but at larger problem sizes the annealing offsets can have a significant impact on performance. The decay observed in success probability in problem sizes larger than 45 implies that insufficient resources were allocated to the tuning procedure, and more than 10,000 samples are needed to tune the offsets. Deriving such an optimal allocation of resources is beyond the scope of this paper. Additional analysis of the final offsets are shown in Appendix B. In real-world applications, it is impractical to exhaustively tune hyperparameters when using heuristics. Typically, either rules-of-thumb or iterative tuning should be used to work in practical timescales. In tuning annealing offsets for the QPU, the existing previous works are two papers from D-Wave Systems that employ two algorithms to change the annealing offsets [11, 12]. In each paper, different magnitudes of samples and success probability improvements are observed. To perform a fair comparison between the various methods, we introduce a performance factor calculated by dividing the number of samples used in the tuning procedure by the improvement factor obtained (the ratio of success probabilities before/after tuning). This performance factor can be interpreted as a measure of resource efficiency, and lower performance factor is better. The best improvement ratio obtained using the CMA-ES procedure introduced in this paper was 12.4, for the configuration of null initial offsets and problem size 45. This yields a factor of 806, which is more than 3 times lower than in the previous best in existing literature. We therefore find our method to be 3 times more efficient than the existing method at best. We also note that both versions of our tuning procedure were more resource efficient (better performance factor) than other methods. A full comparison is shown in Table 2.

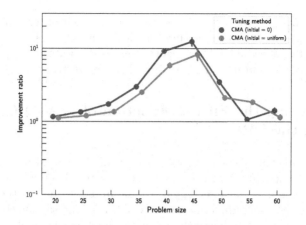

**Fig. 2.** Ratio of mean success probabilities before and after tuning, with different tuner configurations. Means are calculated only using cases where ground states were obtained (as per Fig. 1), and errors were extracted via bootstrapping using 100 data points. Points above the line $10^0$ indicate improvement by using the respective tuning configuration. Peak improvement is observed for problem size 45 with both configurations, with a mean improvement of 12.4 for the null initial offset configuration and a factor of 8.2 for the uniform configuration.

**Table 2.** Table comparing the number of samples used, the success probability improvement ratio, and overall performance factor between the existing annealing offset tuning methods.

| Method | Samples | Improvement | Performance |
|---|---|---|---|
| D-Wave (grid) | $2.5 \cdot 10^6$ | 1000 | 2500 |
| D-Wave (perturb.) | $3.15 \cdot 10^5$ | 1.37 | $2.3 \cdot 10^4$ |
| CMA-ES (uniform) | $10^4$ | 8.2 | 1219 |
| CMA-ES (null) | $10^4$ | 12.4 | **806** |

# 7   Conclusions

In this paper we introduced a novel method for heuristically tuning hyperparameters in existing quantum annealing processors. Using a (1+1)-CMA-ES algorithm to tune annealing offsets, we demonstrate an improvement of up to 12.4 times in probability of obtaining optima, and are able to find better solutions than without tuning. We are able to do this in a model-agnostic way that does not require domain-specific knowledge other than the bounds of the parameters. Additionally, we are able to make efficient use of our QPU samples, and are more than 3 times more efficient than the existing tuning techniques shown in [11,12]. Improvements via tuning were obtained by exploring only a single parameter, the annealing offset parameter. Our results show that it is possible to use classical algorithms to iteratively tune hyperparameters and boost performance of commercially available QPUs, shown here on a test set of MIS problems. This result opens the door to new use cases for classical optimizers, and introduces a new paradigm for hybrid quantum/classical computing. In future work, we will investigate additional tuning algorithms, incorporate more parameters in the tuning process, and use additional problem sets to test the results.

## A   Evaluating the Fitness Function of $(1 + 1)$-CMA-ES Using the QPU

Here we show an example of a single tuning run of $(1+1)$-CMA-ES for a 40 node graph, with the configuration of initial offsets set to all zeroes. As explained in Algorithm 1, we use the mean energy of 100 samples returned by the QPU as the evaluation for the fitness function of the CMA-ES. The sample size of 100 was determined empirically as being the minimum number of samples to determine the mean energy, and is consistent with previous results [4]. In Fig. 3 (left) we show the progression of the CMA-ES routine and the associated fitness function. The tuning shows a clear improvement in mean energy, as shown both in the fitness function and the cumulative minimum of the fitness function. Every time the objective function improves, the respective annealing offsets that were used in that sample set are recorded. The evolution of the annealing offsets for this 40 variable instance is shown in Fig. 3 (right). The final offsets after tuning were then used to test their performance.

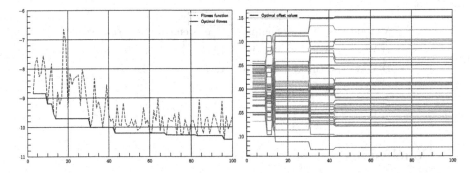

**Fig. 3. Left:** The fitness function evolution (mean energy of 100 samples) is shown as a function of the iteration number in CMA-ES. The red line represents the value of the fitness function at each iteration of CMA-ES, and the blue line is the cumulative minimum, representing the best solutions so far. **Right:** The evolution of the annealing offsets are shown as a function of the iteration number of CMA-ES (updated every time improvement is found by the CMA-ES). (Color figure online)

## B   Analysis of Tuned Annealing Offsets

Here we present the aggregated results of all the annealing offsets post tuning. Figures 4 (left and right) show the final offset values for all problem instances using the CMA-ES routine with initial offsets set to zero and uniform, respectively. We found that the final offset values were not correlated with chain length or success probability. However, we did see a systematic shift in the final offset values with respect to the degree of the node in the graph, and as a function of problem size. In both figures, we see divergent behavior in offsets for very small and very high degree nodes, with consistent stability in the mid-range. The main different between the two figures is the final value of the offsets in this middle region. In Fig. 4 (left), the average offset value rises from 0 at small problems, to roughly .02 for problems with 40 variables, then back down to 0 for the largest problems. There is also a slight increase in average offset value from degree 3 to degree 14, found consistently for all problem sizes. In contrast, Fig. 4 (right) shows that the final offset values were roughly .02 at all problem sizes, apart from the divergent behavior in the extrema of the degree axis. The difference between the two configurations could explain why initial offsets set to zero performed slightly better than the uniform initial offsets. Given the fixed resources of 10,000 samples for calibration per MIS instance, escaping from a local optimum (such as the null initial configuration) becomes increasingly difficult at larger problem sizes, thus degrading the uniform configuration's performance. Other than the results shown here, we were not able to extract any meaningful information with respect to other interesting parameters.

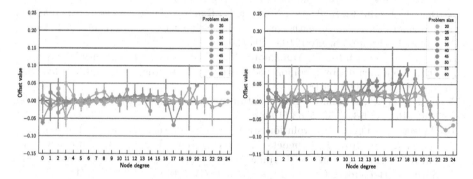

**Fig. 4. Left:** Final offset value as determined by (1+1)-CMA-ES with initial offsets set to zero, as a function of the degree of the logical node in the graph. Colors represent different problem sizes. **Right:** Same as in left, but for initial offsets set uniformly in their allowed range. (Color figure online)

# References

1. King, J., Yarkoni, S., Nevisi, M.M., Hilton, J.P., McGeoch, C.C.: Benchmarking a quantum annealing processor with the time-to-target metric. arXiv:1508.05087 (2015)
2. Bian, Z., Chudak, F., Israel, R.B., Lackey, B., Macready, W.G., Roy, A.: Mapping constrained optimization problems to quantum annealing with application to fault diagnosis. Front. ICT **3**, 14 (2016)

3. Neukart, F., Compostella, G., Seidel, C., von Dollen, D., Yarkoni, S., Parney, B.: Traffic flow optimization using a quantum annealer. Front. ICT **4**, 29 (2017)

4. Raymond, J., Yarkoni, S., Andriyash, E.: Global warming: temperature estimation in annealers. Front. ICT **3**, 23 (2016)

5. Venturelli, D., Marchand, D.J.J., Rojo, G.: Quantum annealing implementation of job-shop scheduling. arXiv:1506.08479 (2015)

6. King, A.D., et al.: Observation of topological phenomena in a programmable lattice of 1,800 qubits. Nature **560**(7719), 456–460 (2018)

7. Yarkoni, S., Plaat, A., Bäck, T.: First results solving arbitrarily structured maximum independent set problems using quantum annealing. In: 2018 IEEE Congress on Evolutionary Computation (CEC), (Rio de Janeiro, Brazil), pp. 1184–1190 (2018)

8. Johnson, M.W., et al.: Quantum annealing with manufactured spins. Nature **473**, 194–198 (2011)

9. Barahona, F.: On the computational complexity of Ising spin glass models. J. Phys. A: Math. Gen. **15**(10), 3241 (1982)

10. Lucas, A.: Ising formulations of many NP problems. Front. Phys. **2**, 5 (2014)

11. Lanting, T., King, A.D., Evert, B., Hoskinson, E.: Experimental demonstration of perturbative anticrossing mitigation using non-uniform driver Hamiltonians. arXiv:1708.03049 (2017)

12. Andriyash, E., Bian, Z., Chudak, F., Drew-Brook, M., King, A.D., Macready, W.G., Roy, A.: Boosting integer factoring performance via quantum annealing offsets https://www.dwavesys.com/resources/publications

13. Hsu, T.-J., Jin, F., Seidel, C., Neukart, F., Raedt, H.D., Michielsen, K.: Quantum annealing with anneal path control: application to 2-sat problems with known energy landscapes. arXiv:1810.00194 (2018)

14. Susa, Y., Yamashiro, Y., Yamamoto, M., Nishimori, H.: Exponential speedup of quantum annealing by inhomogeneous driving of the transverse field. J. Phys. Soc. Jpn. **87**(2), 023002 (2018)

15. Kadowaki, T., Nishimori, H.: Quantum annealing in the transverse ising model. Phys. Rev. E **58**, 5355–5363 (1998)

16. Hansen, N.: The CMA evolution strategy: a comparing review. In: Lozano, J.A., Larrañaga, P., Inza, I., Bengoetxea, E. (eds.) Towards a New Evolutionary Computation: Advances in the Estimation of Distribution Algorithms, pp. 75–102. Springer, Heidelberg (2006). https://doi.org/10.1007/3-540-32494-1_4

17. Igel, C., Suttorp, T., Hansen, N.: A computational efficient covariance matrix update and a (1+1)-CMA for evolution strategies. In: Proceedings of the 8th Annual Conference on Genetic and Evolutionary Computation, GECCO 2006, pp. 453–460. ACM, New York (2006)

18. Auger, A., Hansen, N.: Benchmarking the (1+1)-CMA-ES on the BBOB-2009 Noisy Testbed. In: Proceedings of the 11th Annual Conference Companion on Genetic and Evolutionary Computation Conference: Late Breaking Papers, GECCO 2009, pp. 2467–2472. ACM, New York (2009)

# Foundations and Quantum Technologies

# Quantum Photonic TRNG with Dual Extractor

Mitchell A. Thornton$^{(\boxtimes)}$ and Duncan L. MacFarlane

Southern Methodist University, Dallas, TX 75275, USA
{mitch,dmacfarlane}@smu.edu,
http://lyle.smu.edu/~mitch/qirg

**Abstract.** An enhanced true random number generator (TRNG) architecture based on a photonic entropy source is presented. Photonic TRNGs are known to produce photon sequences at randomly distributed time intervals as well as random superimposed quantum states. We describe a TRNG architecture that takes advantage of both of these sources of entropy with a dual-source extractor function. We show that the amount of harvested entropy exceeds that compared to implementations comprised of only one of these sources. We also describe an implementation of a beam splitter used within the architecture that is suitable for implementation in a photonic integrated circuit that has been simulated and fabricated.

**Keywords:** TRNG · QRNG · Extractor function · Photonic integrated circuit · Hadamard

## 1  Introduction

The lack of high-quality random number sources in otherwise secure systems has been the cause of several well-documented security breaches [4,12,15]. Many devices require a continuous supply of random values to support the implementation of various modern cryptographic methods. It is desirable for the generation of random bit streams to be accomplished at high data rates to support applications such as modern secure high-speed communications. This need is coupled with the added constraint that random values must be of very high quality in terms of their independence and other statistical properties in order to preserve the integrity of encryption protocols. Additionally, high-speed and high-quality random number generators should be as inexpensive, rugged, and reliable as possible when the hosting devices are intended to be mass-produced.

The physical sources used in true random number generators (TRNG) are sometimes referred to as "weakly random sources" since it is practically impossible to measure or observe the source output without adding some degree of determinism, bias, or correlation. For this reason, TRNGs also incorporate extractor functions, or simply "extractors," that transform the output of a weakly random

© Springer Nature Switzerland AG 2019
S. Feld and C. Linnhoff-Popien (Eds.): QTOP 2019, LNCS 11413, pp. 171–182, 2019.
https://doi.org/10.1007/978-3-030-14082-3_15

source into an equally likely and independent string of random bits. Many different weakly random sources have been identified and used in TRNGs, such as those based upon quantum effects, electronic metastability, electronic chaos generation, radioactivity, thermal effects, atmospheric effects, deep space radiators, and others. Because the theory of observing the results of quantum mechanical interactions is based on probabilistic axioms, entropy sources that rely upon the measurement or observation of superimposed quantum state information are considered here and have been used in the past [3]. We utilize such a source with photonic information carriers and show how two independent bit streams may be extracted in an efficient manner with only minor architectural changes required as compared to previous photonic architectures.

## 2    Quantum Photonic TRNG Architecture

In general, TRNGs are comprised of a physical source, an observation or measurement stage, and a post-measurement processing stage known as an "extractor" function, as shown in Fig. 1. The purpose of the extractor is to discard the undesired biases, correlations, and other deterministic components in the source measurements and to transform random values to output values that are as close as possible to being independent and equally likely. From an information theoretic point of view, the goal of the TRNG extractor is to maximize the information entropy in the output values by utilizing as much of the entropy present in the physical source as possible. Furthermore, the extractor function ideally produces values that are independent and uniformly distributed regardless of the native distribution of the physical source observations. For at least these reasons, extraction functions are very important with regard to the quality of TRNG output values. We propose a TRNG implemented in a Quantum Photonic Integrated Circuit (QPIC) using location-encoded (*aka*, "dual-rail") methods for information representation.

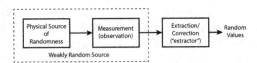

**Fig. 1.** A generic and typical TRNG block diagram including a physical source and extractor function.

We propose an architecture wherein a single photon pump excites a spontaneous parametric down conversion (SPDC) device to generate a heralded single photon source in the form of a signal and idler photon pair. The signal is then applied to a 50-50 beam splitter, used as a Hadamard operator, that drives the two waveguides representing orthogonal basis states, $|0\rangle$ and $|1\rangle$. The idler photon is transmitted in a third waveguide that enables a heralded implementation.

Each of the three waveguides drives a single-photon avalanche diode (SPAD) detector that we denote as SPAD-0, SPAD-1, and SPAD-T. Thus, a random bit stream is produced depending upon which of SPAD-0 and SPAD-1 indicate energy detection that is correlated in time with an active output from SPAD-T. This approach for implementing a TRNG using photonic quantum effects is well-known and variations of this approach are used in commercial devices [2,9–11,14,16]. Figure 2 contains a diagram illustrating this approach.

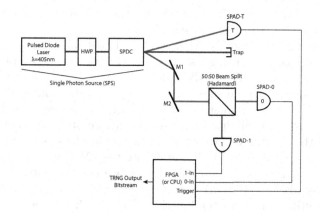

**Fig. 2.** A TRNG comprised of SPS, SPDC, and three SPADs.

As an example, the single photon source (SPS) in Fig. 2 is comprised of a pulsed laser source with wavelength 405 nm serving as a pump and a rotatable half-wave plate (HWP) for adjusting the angle of linear polarization of the pump photon with the optical axis of the spontaneous parametric down converter (SPDC). The down-converted signal and idler photons are at 810 nm wavelength. The topmost beam in Fig. 2 is indicative of the idler and is detected by SPAD-T whereas the bottommost beam produced by the SPDC is indicative of the 810 nm signal photon beam and is path adjusted via mirrors M1 and M2 prior to entering the beam splitter. The beam splitter serves as a location-encoded Hadamard operator thus causing the position state of the signal photon to be equiprobable. The two outputs of the beam splitter and the idler photon are applied to detectors, SPAD-0, SPAD-1, and SPAD-T that supply input signals to an FPGA or CPU. When the FPGA or CPU receives a heralded input due to detection from SPAD-0 or SPAD-1, a randomly generated bit is produced based on which detector was activated.

In practice, a microcontroller (not shown) will provide the proper quenching of the SPAD and will maintain its operation in Geiger mode. The heralding, or triggering modality of the TRNG in Fig. 2 also provides the counts that allow coincidence statistics to ensure the system operated in the quantum regime, in spite of detection imperfections and possible laser drift.

Our proposed architecture is the same as that shown in Fig. 2 with the exception that the FPGA or CPU core is programmed to take advantage of two sources of entropy from the SPS. A physical photon entropy source such as that in Fig. 2 exhibits two different and statistically independent random characteristics. The first is a sequence of measurements based upon whether energy is detected at SPAD-0 or SPAD-1 as just described. The second is the sequence of time intervals between photon detection at either SPAD-0 or SPAD-1. It is irrelevant whether detection occurs at either SPAD-0 or SPAD-1 with respect to the sequence of time intervals separating photon generation. Thus, the time interval sequence is statistically independent with respect to the sequence of generated bits due to SPAD-0 and SPAD-1. Therefore, we consider the SPS as providing two independent entropy sources that are statistically independent.

Recent past results indicate that TRNGs based upon two sources can be superior as compared to a single source TRNG [5,6]. Our two-source approach utilizes two random variables (RV) where one is a Bernoulli distributed RV, $X$, and the other is a time series of sub-Poissonian distributed time intervals denoted as RV $Y$ that originate as characteristics of the same single SPS. The extractor for the Bernoulli distributed RV $X$ with variate $x_i$ is denoted $Ext_2(X)$ and utilizes the outputs of two single-photon avalanche diodes (SPAD) shown as SPAD-0 and SPAD-1 in Fig. 2. We denote the RV $V$ with variate $v_i$ as the extracted value of variate $x_i$. The RV $V$ has two possible outcomes and hence the event space is $\mathbb{F}_2 = \{0, 1\}$.

We also utilize an extractor function, $Ext_r(Y; r)$, for the sub-Poissoinian distributed RV $Y$ that was recently introduced in [18] and referred to herein as "[TT:18]." RV $W$ with variate $w_i$ is extracted from RV $Y$ via the use of $Ext_r(Y; r)$. The variates, $y_i$, of $Y$ as utilized in our proposed TRNG architecture are of the form of a discretized set of time intervals, $\Delta t_i$. The $w_i$ values are radix-$R$ values in the form of a bitstring of length $r$ that have values $w_i \in \mathbb{F}_r = \{0, 1, \cdots, 2^r - 1\}$ where the number of different bitstrings is also $|\mathbb{F}_r| = R$ and where $R = 2^r > 2$. The extractor function $Ext_r(Y; r)$ receives input from SPAD-T in Fig. 2 and is implemented in the FPGA or CPU core that is also shown in Fig. 2.

. Because $X$ and $Y$ are statistically independent and uncorrelated, the overall composite extractor of our TRNG is formed from $Ext_2(X)$ and $Ext_r(Y; r)$ and is denoted as $Ext(X, Y; r) = Ext_2(X)\|Ext_r(Y; r)$ where $\|$ denotes the concatenation operation. The order of concatenation is arbitrary and irrelevant. Generally, any arbitrary permutation of the bitstrings resulting from $Ext(X, Y; r)$ would suffice due to the fact that $V$ and $W$ are equally likely and independent.

**Lemma 1.** *A TRNG with a quantum photonic source as depicted in Fig. 2 and a composite extractor function $Ext(X, Y; r) = Ext_2(X)\|Ext_r(Y; r)$ yields generated values that are uniformly distributed when $Ext_2(X)$ produces a uniformly distributed RV $V$ and $Ext_r(Y)$ produces a uniformly distributed RV $W$.*

*Proof.* The probability that a variate of $V$ is a value in the set $\mathbb{F}_2 = \{0, 1\}$ is $\frac{1}{2}$ since $V$ is uniformly, or in this case, Bernoulli distributed with probability of success $\frac{1}{2}$. Likewise, the probability that a variate of RV $W$ is a value in the set $\mathbb{F}_r$ is $\frac{1}{2^r}$ since $W$ is also uniformly distributed. Since RVs $V$ and $W$ are independent, the probability of the $r + 1$ bit concatenated variate of RV $S$, or $s_i = v_i \| w_i$ is $P[V \| W] = P[X \cap Y] = P[X]P[Y] = \frac{1}{2^{r+1}}$.

## 3   TRNG Theory and Analysis

The detection of photons by either SPAD-0 or SPAD-1 is theoretically modeled as a sequence of events corresponding to observations of a Bernoulli-distributed random variable (RV), $X$, with parameter $p$. In the theoretically ideal case, the Hadamard operator is implemented with a perfect 50:50 beam splitter resulting in the Bernoulli PMF parameter $p$ being exactly $\frac{1}{2}$. However, as discussed in a later section, perfectly ideal beam splitters are not realizable in the laboratory or in manufacturing environments. Thus, an extractor function is used to adjust for practical tolerances in actual beam splitters.

We model the output of a SPAD as the function $f_{SPAD}$ that has a nominal output of 0V. Upon detecting a photon at time $t$, $f_{SPAD}$ produces a rising edge of a short duration pulse where the constant $T_{SPAD}$ represents the short pulsewidth characteristic of the SPAD and $u(t)$ represents a unit step function. The SPAD characteristic behavior as modeled by $f_{SPAD}$ is $f_{SPAD}(t) = u(t) - u(t - T_{SPAD})$ when SPAD-T detects an idler photon at time $t$.

In considering the case of an ideal beamsplitter, the quantum state of the location-encoded photon is maximally superimposed and is of the form $|\phi\rangle = \frac{|0\rangle + |1\rangle}{\sqrt{2}}$ since the parameter $p$ in a Bernoulli probability density function is $\frac{1}{2}$. This results in a photon detection event that is equally likely to happen in either SPAD-0 or SPAD-1 with probability of occurance equal to $\frac{1}{2}$ in response to the production of a signal and idler pair from the SPDC. However, in terms of actual implementations of beam splitters, such an ideal case is never achieved in practice since the devices are fabricated within tolerance levels and may also suffer from other imperfections. Thus the TRNG with an architecture such as that in shown in Fig. 2 is more realistically modeled with the parameter $p$ being of the form $p \neq \frac{1}{2}$. For this reason, we employ the use of the well-known von Neumann extractor function for $Ext_2(X)$ although other previously known extractors such as the Trevisan or Toeplitz Hashing approach may be used depending upon the intended application of the TRNG. For the purpose of this paper, the choice of the particular extractor used for this portion of the circuit is not relevant to the main contribution. Extractor $Ext_2(X)$ produces the extracted sequence of variates $v_i$ from the extracted RV $V$.

The sequence of measured time intervals between photon detection events is representative of a sub-Poissonian process and is denoted by RV $Y$ [1,7,19]. The variates $y_i$ of RV $Y$ are discretized values representing each interval $\Delta t_i$. The detection coincidence window values, $T_{win}$, are chosen and used in the FPGA (or CPU software) in regard to the SPS parameters to ensure that the time intervals between detection pulses from SPAD-T are indeed sub-Poissonian distributed thus minimizing photon number bunching within a measurement interval.

The actual $\Delta t_i \in \mathbb{R}$ time intervals are positive, real, and non-zero. Due to the fact that the TRNG is implemented with a hybrid of photonic, analog, and digital electronic circuitry, the observation and measurement of RV $Y$ results in a discrete positive integer-valued variate, $y_i$, from the interval $y_i \in [n_1, n_2]$. The integer-valued $y_i$ measurement estimates the actual real-valued $\Delta t_i$ value via the relationship $y_i = \lceil \Delta t_i \times \tau \rceil$ where $\tau$ is the clock period of a digital incrementor circuit or counter within the TRNG that counts the number of $\tau$ time intervals that elapse between adjacent photon detection events in time.

Figure 3 contains a plot of the TRNG detector activations. The heralded detector output is indicated on the horizontal axis representing the SPAD-T detector when it detects the presence of an idler photon as shown via a tick mark labeled $t_i$. $t_i$ is the time at which the SPAD-T detects an incident idler photon causing a rising edge of $f_{SPAD}$. The vertical axis is labeled with two events; the detection of a signal photon at either the SPAD-0 or SPAD-1 detector. Each black dot on the plot of Fig. 3 indicates whether the signal photon was detected by the SPAD-0 or SPAD-1 detector. Theoretically, the signal photon is equally likely to be detected at either the SPAD-0 or the SPAD-1 detector since it is placed into maximal and equal superposition due to the Hadamard operator realized as a beam splitter and the resulting extracted $v_i$ value as shown in Fig. 2.

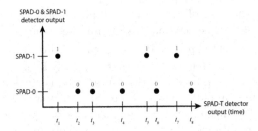

**Fig. 3.** Plot of time sequence of detected photon events by SPAD-0 and SPAD-1.

Figure 3 actually indicates two statistically independent random processes. The first is modeled as RV $V$ and is the equally likely event that the signal photon is detected by either the SPAD-0 or SPAD-1 detector. The second process, denoted as event RV $X$, corresponds to the event that the idler photon is detected by SPAD-T at some time interval $\Delta t_i$ where $\Delta t_i = t_{i+1} - t_i$. Alternatively, the two sets of observations of RVs shown in Fig. 3 can be interpreted as the set of $X$ observations, $\{1, 0, 0, 0, 1, 0, 1, 0\}$ and the set of $Y$ observations $\{y_1, y_2, y_3, y_4, y_5, y_6, y_7\}$ are the discretized values representing $\{\Delta t_1, \Delta t_2, \Delta t_3, \Delta t_4, \Delta t_5, \Delta t_6, \Delta t_7\}$. In terms of information theory, each observation of $X$ and $Y$ yields some amount of self-information of the corresponding extracted values $v_i$ and $w_i$, denoted as $I(v_i)$ and $I(w_i)$. The self-information, in units of bits, that corresponds to the event that RV $A$ is observed to have an outcome of $a_i$ (i.e. $A = a_i$) is given in Eq. 1.

$$I(A = a_i) = -log_2[P(A = a_i)] \tag{1}$$

In the case of the information content of the strings resulting from the composite extractor function, $Ext(X,Y;r) = Ext_2(X)||Ext_r(Y;r)$, the TRNG provides a series of bit strings comprised of substrings, $s_i$, where $s_i$ is the concatenation of the $r$-bit string $w_i$ extracted from the $y_i$ variates using extractor $Ext_r(Y;r) = w_i$, and the corresponding single bit values $v_i$ extracted from variates $x_i$ using the von Neumann extractor $v_i = Ext_2(X)$. Thus, the TRNG produces a series of substrings $s_i$ that are comprised of $r + 1$ bits formed as a concatenation $s_i = v_i||w_i$.

**Lemma 2.** *The concatenated string of $r + 1$ bits, $s_i = v_i||w_i$, contains self information that is the arithmetic sum of the self information of $v_i$ and $w_i$.*

*Proof.* From Lemma 1 it is proven that $s_i$, a variate of RV $S$, is uniformly distributed where $s_i = v_i||w_i$ and where $v_i$ and $w_i$ are each independent variates. Thus $P[s_i] = P[v_i||w_i] = P[v_i \cap w_i] = P[v_i]P[w_i] = P(v_i) \times P(w_i) = (\frac{1}{2})(\frac{1}{2^r}) = \frac{1}{2^{r+1}}$. Using the definition of self-information in Eq. 1:

$$I(s_i) = -log_2[P(v_i) \times P[w_i] = -log_2[P(v_i)] - log_2[P(w_i)] = I(v_i) + I(w_i)$$

For the ideal Hadamard operator in Fig. 3, the self-information due to an observation of RV $X$ is, not surprisingly, one bit in the ideal case. For the RV $W$, the self-information due to the extracted value $w_i$ is based on a substring of size $r$. Since the extracted $w_i$ are ideally uniformly distributed, the self information is:

$$I(w_i) = -log_2[P(w_i)] = -log_2\left[\frac{1}{2^r}\right] = log_2(2^r) = r \tag{2}$$

Information entropy is the expected value of the self-information, $H(A) = E\{I(A)\}$. Thus, for $N_{tot}$ observations of $A$, assuming each $A$ is comprised of $k$ bits, the corresponding information entropy in units of bits is given in Eq. 3.

$$H(A) = E\{I(A)\} = \sum_{i=1}^{k \times N_{tot}} I(A = a_i)P[I(A = a_i)] \tag{3}$$

From probability theory, it is the case that $P[I(A = a_i)] = P[-log_2\{P(A = a_i)\}] = P[A = a_i]$, thus Eq. 3 can be simplified to the well-known form in Eq. 4.

$$H(A) = E\{I(A)\} = \sum_{i=1}^{k \times N_{tot}} P[a_i]log_2(P[a_i]) \tag{4}$$

**Theorem 1.** *A TRNG with a quantum photonic source SPS as depicted in Fig. 2 and a composite extractor function $Ext(X,Y;r) = Ext_2(X)||Ext_r(Y;r)$ harvests more entropy from the SPS source than a TRNG that uses only the extractor $Ext_2(X)$ or only the extractor $Ext_r(Y;r)$.*

*Proof.* RV $X$ and $Y$ are statistically independent since the generation of photon pairs from the SPDC in the TRNG depicted in Fig. 2 occurs probabilistically and before the generated signal photon is placed into a state of superposition by the beamsplitter and subsequently detected by either SPAD-0 or SPAD-1. The $N_{tot}$-length sequence of $\{w_i\}$ is extracted from the $N_{tot}$-length sequence $\{y_i\}$ that are discretized values of $\{\Delta t_1, \Delta t_2, \Delta t_{N_{tot}}\}$. While the $N_{tot}$-length sequence $\{y_i\}$ is a set of discretized sub-Poissonian distributed values of $\{\Delta t_1, \Delta t_2, \Delta t_{N_{tot}}\}$, the corresponding $\{w_i\}$ sequence is a set of uniformly distributed length-$r$ substrings due to extractor $Ext_r(Y; r)$ that are independent with regard to the signal photon being placed into a state of superposition prior to its detection by either SPAD-0 or SPAD-1. Alternatively, the outcome of the extracted $v_i$ value from RV $X$ is due to a fundamental axiom of quantum mechanics that is independent of the time intervals separating signal and idler pair generation from the SPDC.

The maximum amount of entropy available from a sequence of $N_{tot}$ variates $\{v_i\}$ extracted from RV $X$ occurs when the beam splitter is ideal and hence the von Neumann extractor has 100% efficiency and yields $N_{tot}$ random bits when $N_{tot}$ bits are operated over by the extractor. Thus, since each bit is equally likely to be zero or one, the resulting harvested entropy due to $Ext_2(X)$ is calculated on a per bit basis using Eq. 4 resulting in Eq. 5.

$$H(\{v_i\}) = -\sum_{i=1}^{1 \times N_{tot}} P(v_i)log_2[P(v_i)] = -N_{tot}\left(\frac{1}{2}\right)log_2\left(\frac{1}{2}\right) = \frac{N_{tot}}{2} \quad (5)$$

Likewise, the entropy harvested from a sequence of $N_{tot}$ substrings of length $r$, $\{w_i\}$, extracted from RV $Y$ by $Ext_r(Y; r)$ is given by Eq. 4 resulting in Eq. 6.

$$H(\{w_i\}) = -\sum_{i=1}^{r \times N_{tot}} P(w_i)log_2[P(w_i)] = -(r \times N_{tot})\left(\frac{1}{2}\right)log_2\left(\frac{1}{2}\right) = \frac{N_{tot}}{2}(r)$$
$$(6)$$

Finally, the energy harvested from the sequence $\{s_i\}$ of length $N_{tot}$ using the composite extractor $Ext(X, Y; r) = Ext_2(X)||Ext_r(Y; r)$ is given by Eq. 4 resulting in Eq. 7.

$$H(\{s_i\}) = -\sum_{i=1}^{(r+1) \times N_{tot}} P(s_i)log_2[P(s_i)] = -\sum_{i=1}^{(r+1) \times N_{tot}} P(v_i||w_i)log_2[P(v_i||w_i)]$$

$$= -\sum_{i=1}^{(r+1) \times N_{tot}} P(v_i \cap w_i)log_2[P(v_i \cap w_i)] = -\sum_{i=1}^{(r+1) \times N_{tot}} P(v_i)P(w_i)log_2[P(v_i)P(w_i)]$$

$$= -\sum_{i=1}^{(r+1) \times N_{tot}} \left(\frac{1}{2}\right)\left(\frac{1}{2}\right)log_2\left[\left(\frac{1}{2}\right)\left(\frac{1}{2}\right)\right] = [(r+1) \times N_{tot}]\left(\frac{1}{2}\right) = \frac{N_{tot}}{2}(r+1)$$
$$(7)$$

Comparing the entropy $H(\{s_i\})$ in Eq. 7 with $H(\{v_i\})$ in Eq. 5, we can calculate bounds on the value of $r$ to ensure $H(\{s_i\}) > H(\{v_i\})$:

$$\frac{N_{tot}}{2}(r+1) > \frac{N_{tot}}{2} \implies r > 0$$

Thus, as long as $r > 1$, the entropy harvested from timing intervals $w_i$ is larger than that from the state detection $v_i$ and the proof is complete.

## 4   TRNG Implementation

A quantum photonic integrated circuit (QPIC) implementation of the TRNG architecture includes a Hadamard operator, implemented as a beam splitter, internally in the QPIC. The Hadamard operator is realized in a novel way by nanoscale frustrated total internal reflection (FTIR) couplers as shown in Fig. 4 [8,13,17,20]. This component comprises a thin trench that cuts across a waveguide at 45 degrees. This angle promotes total internal reflection (TIR) at the interface of the waveguide and the trench. If, however, the trench is etched thin enough so that a portion of the evanescent field is coupled into the subsequent waveguide, then the TIR is frustrated. To aid in fabrication and reliability, the etched trench is often backfilled by atomic layer deposition (ALD) with a low dielectric constant material. This component therefore allows an integrated realization of the macroscopic beamsplitter realization of a Hadamard gate. Figure 4a depicts a schematic diagram of the arrangement of the integrated waveguides and the nanoscale FTIR coupler. Figure 4b depicts a finite difference time domain (FDTD) simulation of photon tunneling into two perpendicular waveguides using a nanoscale FTIR coupler. A cross section of such a fabricated trench is shown in the scanning electron microscope (SEM) photograph of Fig. 4c, and a scanning electron microscope (SEM) photograph in Fig. 4d of a fabricated and functional nanoscale FTIR coupler.

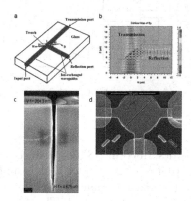

**Fig. 4.** Architecture, FDTD simulation, and scanning electron micro.

The FPGA, or alternatively an embedded CPU core, in Fig. 2 receives output from SPAD-T, SPAD-0, and SPAD-1. Internally the FPGA or CPU processing logic implements the extractor functions and produces a random bit stream as output in addition to other signal conditioning and control functions. Figure 5 depicts a block diagram that partially illustrates the FPGA/CPU functionality of the proposed TRNG depicted in Fig. 2. This functionality is implemented as FPGA processing logic and/or as software in an embedded CPU core.

The parameters of the TRNG denoted as $\tau$ and $R$ in Fig. 5 denote the internal sampling clock period ($\tau$) that is smaller than the measurement coincidence window. The radix value, $R = 2^r$ is used to quantize the timing intervals $\Delta t_i$ that yield the sub-Poissonian distributed variate $y_i \in [n_1, n_2]$. Each time a new idler photon is detected at SPAD-T, the fixed-point timer logic begins a processing cycle. The fixed-point timer logic (FPTL) computes $\Delta t_i = t_{i+1} - t_i$ as a quantized value in the form of an $r$-bit word $y_i$. The FPTL contains an internal incrementor register that is reset by a rising edge on the output of the SPAD-T detector and it is configured as an up-counter that increments every $\tau$ time units as a means to compute $\Delta t_i$. When an idler photon is detected by the SPAD-T, the incrementor first outputs its current discretized count value $y_i$ to the extractor block labeled [TT:18] in Fig. 5, then it resets and begins counting again from zero. The output value of the incrementor is the quantized value of $\Delta t_i$ representing an observation of RV $Y$ for the previous time interval between photon detections with a resolution set by parameter $\tau$. Note that $y_i$ is not necessarily restricted to being $r$ bits in length as is its extracted value, $w_i$.

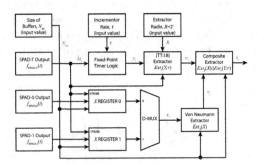

**Fig. 5.** Block diagram of digital processing portion of the TRNG.

The block labeled [TT:18] Extractor implements the extractor function denoted as $Ext_r(Y; r)$ as described in [18]. It produces an $r$-bit value $w_i$ whose value is in the set $\mathbb{F}_r = \{0, 1, \cdots, 2^{r-1}\}$ and that is uniformly distributed and produced from the input quantized $y_i$ values derived from corresponding $\Delta t_i$ values that are both sub-Poissonian distributed. This block also contains a buffer of a length suitable length to store $N_{tot}$ different $y_i$ and $w_i$ sample values. The [TT:18] extractor block receives $N_{tot}$ quantized $y_i$ values, applies them to the $Ext_r(Y; r)$ function, and yields $N_{tot}$ different $w_i$ output values using the methodology described in [18]. The outputs of SPAD-0 and SPAD-1 are registered into single bit registers labeled "X Register 0" and "X Register 1" depending upon which SPAD outputs a pulse. The registers have values that are strobed in only when the SPAD-T rising edge occurs thus ensuring that the SPAD activations are due to an actual produced signal/idler pair from the SPDC versus some other spurious or extraneous detections. After an appropriate delay in the SPAD-T output signal (the delay element is not shown in Fig. 5) the demultiplexer logic outputs either a "0" or "1" electronic bit into the von Neumann extractor logic circuit depending on which of SPAD-0 or SPAD-1 was activated.

The von Neumann $Ext_2(X)$ extractor logic contains an internal buffer in the form of a serial input shift register that is also of length $N_{tot}$. When $N_{tot}$ bits representing variates of RV $X$ have been accumulated, the von Neumann extractor function evaluates thus ensuring that the samples, $v_i$, of RV $V$ are indeed equiprobable. The block labeled "Composite Extractor" $Ext(X, Y; r) = Ext_2(X)||Ext_r(Y; r)$ receives the $N_{tot}$ extracted bitstrings of length $r$, denoted as variates $w_i$, from the [18] $Ext_r(Y; r)$ block and the corresponding $N_{tot}$ extracted bits from the von Neumann $Ext_2(X)$ extractor block. It then concatenates each $r$-bit value $w_i$ with each matching single bit extracted value $v_i$ and ideally outputs $N_{tot}$ concatenated bit strings, $s_i = v_i||w_i$ each of length $r+1$, as the TRNG output. Although we describe the composite extractor function $Ext(X, Y; r) = Ext_2(X)||Ext_r(Y; r)$ as performing concatenation resulting in random bit substrings of the form $s_i = v_i||w_i$, it is actually the case that the random bit $v_i$, can be inserted into any arbitrary location within the random bit string $s_i$ without any degradation in terms of TRNG output quality.

## 5    Conclusion

A new TRNG architecture is presented that has enhanced throughput through the use of a new two-source extractor function that utilizes two sources of physical entropy from a single photonic source; a random sequence of time intervals and the randomness present in measurements of a superimposed quantum state. We also propose a novel structure that has been designed, fabricated, and tested in the QPIC of Fig. 4 that serves as the Hadamard operator for the purpose of transforming the quantum state of a signal photon into a superimposed state prior to its detection by either SPAD-0 or SPAD-1. This TRNG is suitable for implementation on a hybrid integrated circuit containing both photonic and electronic processing. The new extractor functions are implemented either within an on-chip FPGA or embedded electronic CPU core.

## References

1. Arnoldus, H.F., Nienhuis, G.: Conditions for sub-poissonian photon statistics and squeezed states in resonance fluorescence. Optica Acta **30**(11), 1573–1585 (1983)
2. Baetoniu, C.: Method and Apparatus for True Random Number Generation. U.S. Patent 7,389,316, 17 June 2008
3. Dulz, W., Dulz, G., Hildebrandt, E., Schmitzer, H. (inventors): Method for Generating a Random Number on a Quantum-Mechanical Basis and Random Number Generator. U.S. Patent 6,609,139, 19 August 2003

4. Dorrendorf, L., Gutterman, Z., Pinkas, B.: Cryptanalysis of the random number generator of the Windows operating system. ACM Trans. Inf. Syst. Secur. 13(1), Article no. 10 (2009)
5. Chattopadhyay, E.: Explicit two-source extractors and more. Ph.D. dissertation, The University of Texas at Austin, May 2016
6. Chattopadhyay, E., Zuckerman, D.: Explicit two-source extractors and resilient functions. In: Proceedings of ACM Symposium of the Theory of Computing (STOC), pp. 670–683, June 2016
7. Fox, M.: Quantum Optics: An Introduction. Oxford University Press, Oxford (2006). ISBN 13-978-0-19-856673-1
8. Huntoon, N.R., Christensen, M.P., MacFarlane, D.L., Evans, G.A., Yeh, C.S.: Integrated photonic coupler based on frustrated total internal reflection. Appl. Opt. **47**, 5682 (2008)
9. Hart, J.D., Terashima, Y., Uchida, A., Baumgartner, G.B., Murphy, T.E., Roy, R.: Recommendations and illustrations for the evaluation of photonic random number generators. APL Photonics **2**, 090901 (2017). https://doi.org/10.1063/1.5000056
10. ID Quantique, SA: Quantis Random Number Generator. http://certesnetworks. com/pdf/alliance-solutions/QNRG-When-Randomness-Can-Not-Be-Left-To-Cha nce.pdf. Accessed 16 Nov 2018
11. Jennewein, T., Achleitner, U., Weihs, G., Weinfurter, H., Zeilinger, A.: A fast and compact quantum random number generator. Rev. Sci. Instrum. **71**(4), 1675 (2000)
12. Koerner, B.: Russians Engineer a Brilliant Slot Machine Cheat—And Casinos Have No Fix. Wired Magazine, 06 February 2017
13. Liu, K., Huang, H., Mu, S.X., Lin, H., MacFarlane, D.L.: Ultra-compact three-port trench-based photonic couplers in ion-exchanged glass waveguides. Opt. Commun. **309**, 307–312 (2013)
14. qutools GmbH: Quantum Random Number Generator. Product datasheet (2010). http://www.qutools.com/products/quRNG/quRNG_datasheet.pdf. Accessed 9 June 2018
15. Shumow, D., Ferguson, N.: On the Possibility of a Back Door in the NIST SP800-90 Dual EC. http://rump2007.cr.yp.to/15-shumow.pdf
16. Stipcevic, M.: QBG121 Quantum Random Number Generator, Datasheet, v. 20060328. http://www.irb.hr/users/stipcevi/index.html. Accessed 9 June 2018
17. Sultana, N., Zhou, W., LaFave Jr., T.P., MacFarlane, D.L.: HBr based ICP etching of high aspect ratio nanoscale trenches in InP: considerations for photonic applications. J. Vac. Sci. Technol., B **27**, 2351 (2009)
18. Thornton, M.A., Thornton, M.A.: Multiple-valued random digit extraction. In: Proceedings of IEEE International Symposium on Multiple-Valued Logic (ISMVL), pp. 162–167, May 2018
19. Zou, X., Mandel, L.: Photon-antibunching and sub-Poissonian photon statistics. Phys. Rev. A **41**(1), 475–476 (1990)
20. Zhou, W., Sultana, N., MacFarlane, D.L.: HBr-based inductively coupled plasma etching of high aspect ratio nanoscale trenches in GaInAsP/InP. J. Vac. Sci. Technol., B **26**, 1896 (2008)

# Secure Quantum Data Communications Using Classical Keying Material

Michel Barbeau$^{(\boxtimes)}$ (iD)

School of Computer Science, Carleton University, 1125 Colonel By Drive, Ottawa, ON K1S 5B6, Canada
barbeau@scs.carleton.ca

**Abstract.** We put together recent results on quantum cryptography and random generation of quantum operators. In a synthetic way, we review the work on quantum message authentication by Aharonov et al. and Broadbent and Wainewright. We also outline the work on asymmetric and symmetric-key encryption of Alagic et al. and St-Jules. Quantum operators, i.e., Cliffords and Paulis, play the role of keys in their work. Using the work of Koenig and Smolin on the generation of symplectics, we examine how classical key material, i.e., classical bits, can mapped to quantum operators. We propose a classical key mapping to quantum operators used in quantum cryptography. It is a classical cryptography interface to quantum data cryptography. The main advantage is that classical random key generation techniques can be used to produce keys for quantum data cryptography.

**Keywords:** Quantum information · Quantum data ·
Quantum communications · Quantum cryptography ·
Clifford · Pauli · Symplectic

## 1 Introduction

This paper is about securing quantum data during communications. Two security issues are addressed: authentication and confidentiality. Authentication aims at demonstrating that quantum data has not been altered by an adversary during the transit between a source and a destination. Authentication is accomplished by a data signature mechanism. The goal of confidentiality is to insure the secrecy of data during the transport from a source to a destination. Confidentiality is achieved by encrypting data.

We review three schemes for secure quantum message communication achieving authentication and confidentiality. The three schemes act on quantum data and result from works on cryptographic techniques for quantum message authentication by Aharonov et al. [1] and Broadbent and Wainewright [7,16] and asymmetric and symmetric-key encryption by Alagic et al. [2] and St-Jules [15]. These techniques use quantum operators as cryptographic primitives, i.e., Cliffords and

© Springer Nature Switzerland AG 2019
S. Feld and C. Linnhoff-Popien (Eds.): QTOP 2019, LNCS 11413, pp. 183–195, 2019.
https://doi.org/10.1007/978-3-030-14082-3_16

Paulis. In this paper, we pay a special attention to the problem of mapping classical keying material, i.e., classical bits, to the quantum operators used in quantum cryptograhy. We revisit the algorithm of Koenig and Smolin [13] for mapping an integer to a symplectic operator, which can be used to produce a Clifford. We provide an alternative implementation of the algorithm in MATLAB. We discuss the application of the algorithm for generating quantum operators used in quantum cryptography from classical keying material.

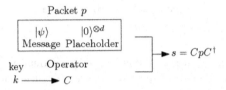

**Fig. 1.** Sender-side authentication procedure for a $m$-qubit message $|\psi\rangle$.

The sender-side authentication procedure is schematically outlined in Fig. 1. A packet $p$ comprises a $m$-qubit message $|\psi\rangle$ and a placeholder for a quantum signature $|0\rangle^{\otimes d}$, where $d$ is a security parameter. A quantum operator $C$ is used to transform the packet into a quantum state $s$. This transformation, $CpC^\dagger$, computes the quantum signature into the placeholder. Figure 2, the receiver gets a $m + d$ qubit state $s' = |s'_1, \ldots, s'_m, s'_{m+1}, \ldots, s'_{m+d}\rangle$. It applies the inverse transformation using the same quantum operator $C$ to obtain the quantum state $r$, i.e., $r = C^\dagger s' C$. When the quantum state has not been altered, $s'$ is equal to $s$. Measurement of qubits $|r_{m+1}, \ldots, r_{m+d}\rangle$ should yield zero. The original quantum message is in qubits $|r_1, \ldots, r_m\rangle$. Otherwise, quantum authentication failed and the message should be rejected. The quantum operators are indexed. We leverage this fact to propose a mapping between a classical $k$ to a quantum operator $C$. Hence, classical key management techniques can be used for quantum data.

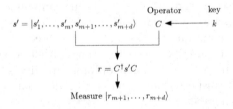

**Fig. 2.** Receiver-side authentication procedure for a $m$-qubit message.

In Sect. 2, we introduce background concepts and more particularly the notion of Clifford group, on which is based the following content. In Sect. 3, we examine quantum message authentication. In Sects. 4 and 5, we outline symmetric-key and asymmetric-key quantum encryption. The use of classical keying material is discussed in Sect. 6. We conclude in Sect. 7. This paper requires familiarity with group theory, namely, the concepts of group, subgroup, left coset, Lagrange theorem [11] and symplectic [8]. In this paper, $\mathbb{F}_2$ represents the two-element field made of 0 and 1. When operands are in $\mathbb{F}_2$, additions (+) are modulo two. This is also equivalent to a logical exclusive or $\oplus$.

# 2 Background and Notation

## 2.1 Quantum Computing Notation

A $n$-qubit quregister is denoted as $|\psi\rangle$. The set of all unitaries of dimension $n$, i.e., acting on $n$-qubit registers, is denoted as $U(2^n)$. $U(1)$ denotes the set all complex numbers modulus one, that is $\{x : |x| = 1\}$. $U^\dagger$ is called the Hermitian conjugate of $U$ (transposition and complex conjugation). The tensor product of two quantum states $|\phi\rangle$ and $|\psi\rangle$ is denoted as $|\phi\rangle \otimes |\psi\rangle$. Given the Pauli operators

$$I = \begin{pmatrix} 1 & 0 \\ 0 & 1 \end{pmatrix}, X = \begin{pmatrix} 0 & 1 \\ 1 & 0 \end{pmatrix}, Y = \begin{pmatrix} 0 & -j \\ j & 0 \end{pmatrix} \text{ and } Z = \begin{pmatrix} 1 & 0 \\ 0 & -1 \end{pmatrix}$$

the Pauli matrices $\mathcal{P}(n)$ on $n$ qubits is the set of matrices of dimension $2^n \times 2^n$ with the general form defined by the tensor product $P_1 \otimes \cdots \otimes P_n$ where $P_1, \ldots, P_n$ are in the set $\{I, X, Y, Z\}$. Because there are four choices for each dimension, $I, X, Y$ or $Z$, the cardinal of $\mathcal{P}(n)$ is

$$|\mathcal{P}(n)| = 4^n = 2^{2n}. \tag{1}$$

## 2.2 Group Theory Concepts

**Pauli Group.** The Pauli group $G_n$ is the set of matrices $\mathcal{P}(n)$ with the factors $\pm 1$ and $\pm j$ [14]. Note that $j$ denotes the value $\sqrt{-1}$. For example, $G_1$ is equal to $\{\pm I, \pm jI, \pm X, \pm jX, \pm Y, \pm jY, \pm Z, \pm jZ\}$. The order of $G_n$ is $4^{n+1}$.

**Symplectic Group.** Let $\mathbb{F}_2$ denote the two-element field comprising 0 and 1. Let $\mathbb{F}_2^{2n}$ denote the field of column vectors of size $2n$ with entries in the field $\mathbb{F}_2$. Let $\Lambda(n)$ be the $2n \times 2n$ block diagonal matrix of the form

$$\begin{pmatrix} \begin{pmatrix} 0 & 1 \\ 1 & 0 \end{pmatrix} & 0 & \cdots & 0 \\ 0 & \begin{pmatrix} 0 & 1 \\ 1 & 0 \end{pmatrix} & \cdots & 0 \\ \vdots & \vdots & \ddots & \vdots \\ 0 & 0 & \cdots & \begin{pmatrix} 0 & 1 \\ 1 & 0 \end{pmatrix} \end{pmatrix}.$$

The symplectic group $Sp(2n)$ on $\mathbb{F}_2^{2n}$ is the set of $2n \times 2n$ matrices $S$ with entries in the field $\mathbb{F}_2$ such that

$$S\Lambda(n)S^T = \Lambda(n)$$

that is, the conjugation of $\Lambda(n)$ using $S$ yields $\Lambda(n)$ [8]. This is termed *preservation of the symplectic inner product* (it also means preservation of commutation). Matrix $\Lambda(n)$ is called the skew-matrix (or antisymmetric or antimetric matrix). The group has been named after the Greek adjective *symplectic* that means *complex*.

**Clifford Group.** Let $\mathcal{P}(n)^*$ denote the set of Pauli matrices on $n$ qubits, excluding the identity matrix, i.e., $\mathcal{P}(n)^*$ is equal to $\mathcal{P}(n)\backslash\{I^{\otimes n}\}$. The Clifford group on $n$ qubits is defined as the set [4,5]

$$C(n) = \{C \in U(2^n) : P \in \pm\mathcal{P}(n)^* \Rightarrow CPC^\dagger \in \pm\mathcal{P}(n)^*\}/U(1). \qquad (2)$$

Firstly, the term $C \in U(2^n)$ says that a Clifford matrix $C$ is a unitary of dimension $2^n \times 2^n$, i.e., it is acting on $n$-qubit registers. On an element $P$ of $\pm\mathcal{P}(n)^*$, the expression $CPC^\dagger$, is called the conjugation of $P$ by matrix $C$. Since the implication indicates that the result is in the set $\pm\mathcal{P}(n)^*$, the set of matrices $\pm\mathcal{P}(n)^*$ is closed under the conjugation by a matrix $C$ in the Clifford group $C(n)$. The term $U(1)$ denotes the set all complex numbers modulus one, that is $\{e^{j\theta} : \theta \in \mathbb{R}\}$. The modulo $U(1)$ suffix, i.e., $/U(1)$, means that two Cliffords $C$ and $D$, that differ only by a complex-number modulus one factor, are equivalent. That is, if $C = \alpha D$, where $\alpha \in U(1)$, then $C \equiv D$. Given a Pauli matrix $P$, the conjugation $CPC^\dagger$ is equal to $(e^{j\theta}C)P(e^{j\theta}C)^\dagger$. The order of the Clifford group $C(n)$ is determined by the following equation [13]

$$|C(n)| = \prod_{j=1}^{n} 2(4^j - 1)4^j = 2^{n^2+2n} \cdot \prod_{j=1}^{n} (4^j - 1). \qquad (3)$$

There is a relationship between the Clifford, Pauli and symplectic groups $C(n)$, $G_n$ and $Sp(2n)$ that we use. That is, there is an equivalence between the set $C(n)/G_n$ and the group $Sp(2n)$. This is interpreted as, given a symplectic $S$ in group $Sp(2n)$ and a Pauli $P$ in group $G_n$, there is a corresponding Clifford $C$ in $C(n)$ equals to the composition $P \cdot S$.

**Corollary 1.** *The order of* $Sp(2n)$ *is* $2^{n^2} \cdot \prod_{j=1}^{n} (4^j - 1)$.

*Proof.* According to Lagrange's theorem, the set $C(n)/(G_n/U(1))$ divides the group $C(n)$ into equal size partitions, i.e., the corresponding left cosets are all of equal size $|C(n)|/|G_n/U(1)|$, using Eq. 3 for $|C(n)|$ and Eq. 1 for $|G_n/U(1)|$, ignoring the phase for elements in $G_n$.

# 3    Quantum Message Authentication

The attack model is the application by the adversary of a Pauli operator, on a qubit state. That is, the perpetration of an attack corresponds to the application of an arbitrary member of $\mathcal{P}(n)$ on a quantum message of size $n$ qubits.

Quantum message authentication uses a cryptographic technique originally introduced by Aharonov et al. [1], with follow up work by Broadbent and Wainewright [7,16]. Quantum message authentication uses the concept of Clifford group, see Sect. 2.2. It involves a message of size $m$ represented as a quantum state $|\psi\rangle$ and a security parameter $d$ (a positive integer). Let $n$ be equal to $m$ plus $d$. A Clifford operator $C$ in group $C(n)$ is selected. The protocol is as follows:

1. Given the message represented by the quantum state $|\psi\rangle$, the source prepares the packet $p$ equal to $|\psi\rangle\langle\psi| \otimes |0\rangle\langle0|^{\otimes d}$. In the tensor product, the left term represents the payload while the right term acts as a placeholder for a signature. The source computes $s = CpC^{\dagger}$.
2. The source sends $s$.
3. While the quantum state $s$ is transferred from the source to the destination, it may be tampered by an adversary. The outcome of the transfer is quantum state $s'$.
4. The destination receives $s'$.
5. The destination computes $r = C^{\dagger}s'C$.
6. The destination measures the $d$ rightmost qubits in quantum state $r$. They correspond to the signature. Corresponding to the payload, the $m$ leftmost message qubits in $r$ are considered valid when the outcome of measurement is $|0\rangle^{\otimes d}$. Otherwise, they are considered invalid.

**Lemma 1.** *If $s'$ is equal to $s$, then $r$ is equal to $p$.*

*Proof.* The destination applies the source's inverse transformation, therefore nullifying its effect. That is, $r$ is equal to $C^{\dagger}s'C$ equal to $C^{\dagger}sC$, which is equal to $C^{\dagger}CpC^{\dagger}C$. Because $C^{\dagger}$ is the inverse of $C$, the later simplifies to $p$.

The transfer of quantum data from a source to a destination can be done using teleportation [3] or quantum repeaters [6]. Element of the group $C(n)$ can be indexed. The index can be interpreted as a classical key. The classical key to Clifford operator mapping is further discussed in Sect. 6.

# 4    Symmetric-Key Quantum Message Encryption

The asymmetric and symmetric-key quantum cryptography techniques used in this and following sections are the original work of Alagic et al. [2] and St-Jules [15]. The symmetric encryption procedure is based on a shared classical key $k$ and a one-way hash function $f$. The hash function takes as arguments the key $k$, a random parameter $i$ and returns a value in $1, \ldots, |G_{2n}|$. It is used to index elements in group $G_{2n}$. Value $i$ plays the role of initialization vector. A repeated usage of key $k$ does not allow an adversary to infer a relationship between encrypted messages. Let $n$, a positive integer, be a security parameter. The set $\{0,1\}^n$ contains all binary strings of length $n$ bits. Key $k$ is of size $n$ bits and $i$ is of size $2n$ bits. Function $f$ outputs a hash value of size $2n$. Hence, its signature is $f : \{0,1\}^n \cdot \{0,1\}^{2n} \to \{0,1\}^{2n}$. Function $f$ is semantically secure, or a one-way hash function. An output of $f$ does not reveal any information about its input.

Let $|\psi\rangle$ be a quantum state of dimension $2n$ representing a message. Encryption is done as follows:

1. A classical binary key $k$ is randomly chosen, with uniform distribution, from the set $\{0,1\}^n$ of binary strings of length $n$ bits.
2. Value $i$ is randomly chosen, with uniform distribution, from the set $\{0,1\}^{2n}$ of binary strings of length $2n$ bits.
3. The source computes the tensor product $|s\rangle = |i\rangle \otimes P_{f(k,i)}|\psi\rangle$, where $P_{f(k,i)}$ is the $f(k,i)$-th Pauli operator of the Pauli group $G_{2n}$.
4. The source sends $|s\rangle$.

The cipher text is decrypted as follows:

1. The destination receives $|s'\rangle$, a quantum state of dimension $2n$ representing an encrypted message.
2. The first $2n$ qubits of $|s'\rangle$ are measured. The result is assigned to $i'$, the initialization vector.
3. Let $P_{f(k,i')}$ be the dimension $n$ Pauli operator determined by the index $f(k,i')$. That is, the $f(k,i')$-th Pauli operator of the Pauli group $G_{2n}$. To obtain the plain text, represented as $|\psi\rangle$, the Pauli operator $P_{f(k,i')}$ is applied to the last $2n$ qubits of $|s'\rangle$, that is, $|\alpha\rangle = P_{f(k,i')}|s'_{2n+1,\ldots,4n}\rangle$.

**Lemma 2.** *If $|s'\rangle$ is equal to $|s\rangle$, then $|\alpha\rangle$ is equal to $|\psi\rangle$.*

*Proof.* The destination applies the source's inverse transformation, therefore nullifying its effect. Because $|s'\rangle$ is equal to $|s\rangle$, $i'$ is equal to $i$ and $P_{f(k,i')}$ is equal to $P_{f(k,i)}$. The Pauli operators are self-inverse, that is, $P_{f(k,i')}P_{f(k,i)} = I$. Hence, we have $P_{f(k,i')}|\alpha_{2n+1,\ldots,4n}\rangle$ equals to $P_{f(k,i')}P_{f(k,i)}|\psi\rangle$, which is equal to $|\psi\rangle$.

## 5   Asymmetric-Key Quantum Message Encryption

The encryption procedure uses a classical public key in $\{0,1\}^n$, the set of bit strings of length $n$. The decryption procedure uses a classical private key in $\{0,1\}^n$. An indexed permutation function $f_i$ is used, with signature $f_i : \{0,1\}^n \to \{0,1\}^n$. The indexed function $f_i$ must have the one-way and *trapdoor* properties [10]. The latter means that, given an output $f_i(x)$ and a trapdoor value $t$, the input $x$ can be determined in polynomial time. Using function $f_i$, a function $h$ that has the hard-core property is constructed to generate pseudo-random bit strings [12]. Function $h$ returns a value in $1, \ldots, |G_{2n}|$, used to index elements in group $G_{2n}$. The hardcore property means that given $f_i(x)$, it is difficult to determine $h(x)$. The index $i$ plays the role of public key. The trapdoor value $t$ plays the role of private key. It is used to resolve the inverse permutation of $f_i(x)$, denoted $g_i(x)$.

Let $|\psi\rangle$ be a quantum state representing a message of dimension $n$. Encryption is done as follows:

1. A value $d$ is selected at random from the domain support of function $f_i$, that is, $d$ is selected such that $f_i(d)$ is not null.
2. Load into variable $r$ the hardcore of $d$, i.e., $h(d)$.
3. The source computes the tensor product $|s\rangle = |f_i(d)\rangle \otimes P_r|\psi\rangle$, where $P_r$ is $r$-th Pauli operator of the Pauli group $G_{2n}$.
4. The source sends $|s\rangle$.

Decryption is done as follows:

1. The destination receives $|s'\rangle$, a quantum state of dimension $2n$ representing an encrypted message.
2. The first $n$ qubits of $|s'\rangle$ are measured. The resulting classical bit string is denoted as $v$.
3. Using trapdoor $t$, apply the inverse permutation $g_i$ of $f_i$ to $v$. The result is stored in $d'$.
4. Calculate the hardcore of $d'$ to obtain $r'$, that is, $r'$ is equal to $h(d')$.
5. To obtain the plain text, apply the Pauli operator $P_{r'}$ to the $n$ rightmost qubits of $|s'\rangle$, that is,

$$|\alpha\rangle = P_{r'}|s'_{n+1,\ldots,2n}\rangle.$$

**Lemma 3.** *If $|s'\rangle$ is equal to $|s\rangle$, then $|\alpha\rangle$ is equal to $|\psi\rangle$.*

*Proof.* The destination applies the source's inverse transformation, therefore nullifying its effect. Because $|s'\rangle$ is equal to $|s\rangle$, we have that $r' = h(d') = h(g_i(v)) = r$ and $P_{r'}$ is equal to $P_r$. The Pauli operators are self-inverse, that is, $P_{r'}P_r = I$. Hence, we have $P_{r'}|\alpha_{n+1,\ldots,2n}\rangle = P_{r'}P_r|\psi\rangle = |\psi\rangle$.

## 6   Mapping Classical Keys to Quantum Operators

Quantum message authentication, Sect. 3, requires a key $k$ to determine a Clifford operator $C$ in $C(n)$. To accomplish this operation, the equivalence, introduced in Sect. 2.2, between the set $C(n)/G_n$ and group $Sp(2n)$ is exploited. That is, a symplectic $S$ in group $Sp(2n)$ and a Pauli $P$ in group $G_n$ decide a corresponding Clifford $C$ in $C(n)$ as the composition $P \cdot S$. According to this scheme, the key $k$ becomes a pair $(k_1, k_2)$, with $0 \le k_1 < |Sp(2n)|$ and $0 \le k_2 < |G_n|$. Using $k_2$, a Pauli can easily be generated decomposing it into $n$ numbers, in $0, 1, 2$ and $3$, each determining the choice of one of the four matrices $I$, $X$, $Y$ and $Z$ then taking their tensor product. Mapping $k_1$ to a symplectic is more involving and discussed hereafter. The symplectic algorithm builds on the concept of subgroup algorithm symplectic inner product and transvection. We review and give our own interpretation of the original work of Koenig and Smolin by [13] on that topic.

Asymmetric-key and symmetric-key message encryption, Sects. 4 and 5, require calculation of the $f(k, i)$-th member of the Pauli group $G_{2n}$. Again, for this a Pauli is produced decomposing $f(k, i)$ into $2n$ numbers, in $0, 1, 2$ and $3$, each determining the choice of one of the four matrices $I$, $X$, $Y$ and $Z$ then taking their tensor product.

**Subgroup Algorithm.** The mapping is based on the subgroup algorithm [9]. The algorithm leverages the following structure. Let $G$ denote a finite group. Let $i$ and $r$ be integers, with $0 \leq i < r$. Let us assume the following sequence of $r$ subgroups $G_r \subset \cdots \subset G_1 \subset G_0 = G$. Any element of a left coset is a *representative* of that coset. A set of representatives for all the left cosets is called a complete set of left coset representatives (members from each left coset). For $i$ in $0, \ldots, r - 1$, let $C_i$ be a complete set of left coset representatives of $G_{i+1}$ in $G_i$. The group $G$ can be equivalently represented as the Cartesian product

$$C_0 \times C_1 \times \cdots \times C_{r-1} \times G_r.$$

In this Cartesian product, every member $g$ in $G$ has a unique equivalent representation

$$(g_0, g_1, \ldots, g_{r-1}, g_r). \tag{4}$$

**Lemma 4.** *The product $C_i \times G_{i+1}$ is equivalent to the subgroup $G_i$, i.e., under the isomorphism $\phi : C_i \times G_{i+1} \to G_i$ with $\phi(c_i, g_{i+1}) = c_i \cdot g_{i+1}$.*

*Proof.* $C_i$ is a complete set of left coset representatives of $G_{i+1}$ in $G_i$. Given an element $c_i$ in the set of representatives $C_i$, the product $c_i G_{i+1}$ is a left coset of subgroup $G_i$. Since every left coset of subgroup $G_i$ is represented by an element $c_i$ in the set of representatives $C_i$, the product $c_i G_{i+1}$ yields that left coset and the product $C_i \times G_{i+1}$ is equivalent to the subgroup $G_i$. That is to say, given a pair $(c_i, g_{i+1})$ in $C_i \times G_{i+1}$, the composition $c_i \cdot g_{i+1}$ is in $G_i$. Conversely, given an element $g_i$ in subgroup $G_i$, there is a pair $(c_i, g_{i+1})$ in $C_i \times G_{i+1}$ such that the composition $c_i \cdot g_{i+1}$ is equal to $g_i$.

**Corollary 2.** *Given a tuple $(g_0, g_1, \cdots, g_{r-1}, g_r)$ in the Cartesian product $C_0 \times C_1 \times \cdots \times C_{r-1} \times G_r$, the composition $g_0 \cdot g_1 \cdots g_{r-1} \cdot g_r$ is in $G$. Conversely, given an element $g$ in group $G$, there is a tuple $(g_0, g_1, \cdots, g_{r-1}, g_r)$ in the Cartesian product $C_0 \times C_1 \times \cdots \times C_{r-1} \times G_r$ such that the composition $g_0 \cdot g_1 \cdots g_{r-1} \cdot g_r$ is equal to $g$.*

In the tuple of Eq. 4, the interesting property is that when the $g_i$'s and $g_r$ are uniformly selected at random, the resulting composition $g$ is uniformly selected at random.

**Symplectic Inner Product and Transvection.** Let $v$ and $w$ be two column vectors in $\mathbb{F}_2^{2n}$. Their symplectic inner product is defined as

$$\langle v, w \rangle = v^T \Lambda(n) w. \tag{5}$$

Note that Eq. 5 can be equivalently formulated as the sum of products

$$v_1 \cdot w_2 + v_2 \cdot w_1 + v_3 \cdot w_4 + v_4 \cdot w_3 + \cdots + v_{2n-1} \cdot w_{2n} + v_{2n} \cdot w_{2n-1}.$$

Let $h$ be a column vector in $\mathbb{F}_2^{2n}$. The vector $h$ is used to define the *symplectic transvection* $Z_h$, with domain and co-domain $\mathbb{F}_2^{2n}$, i.e., $Z_h : \mathbb{F}_2^{2n} \mapsto \mathbb{F}_2^{2n}$. Its application on a column vector $v$ in $\mathbb{F}_2^{2n}$ is defined as $Z_h v = v + \langle v, h \rangle h$. Note that $\langle v, h \rangle$ is a symplectic inner product.

**Lemma 5 (Lemma 2 in [13]).** *Given two non-zero vectors $x$ and $y$ in $\mathbb{F}_2^{2n} \backslash \{0\}$, we have that either $y = Z_h x$ for a $h \in \mathbb{F}_2^{2n}$ or $y = Z_{h_2} Z_{h_1} x$ for $h_1, h_2 \in \mathbb{F}_2^{2n}$. That is to say, vector $x$ can be mapped to vector $y$ using one or two transvections.*

*Proof.* The proof is constructive and yields the transvections. If $x$ and $y$ are equal, then $h_1$ and $h_2$ are zero vectors. If the symplectic inner product $\langle x, y \rangle$ is equal to one, then $h_1 = x \oplus y$ and $h_2$ is a zero vector. If the symplectic inner product $\langle x, y \rangle$ is equal to zero, then find a column vector $z$ in $\mathbb{F}_2^{2n}$ such that $\langle x, z \rangle = \langle y, z \rangle = 1$. Firstly, try to find an index $j \in 2, 4, \cdots, 2n$ where $(x_{j-1}, x_j) \neq (0,0)$ and $(y_{j-1}, y_j) \neq (0,0)$. Find values for a pair $(z_{j-1}, z_j)$ such that $x_{j-1} z_j \oplus x_j z_{j-1} = y_{j-1} z_j \oplus y_j z_{j-1} = 1$, with all other elements of $z$ set to null. Note here that we have both $\langle x, z \rangle = 1$ and $\langle y, z \rangle = 1$. Furthermore, making $h_1 = x \oplus z$ and $h_2 = y \oplus z$, so are $\langle x, h_1 \rangle = 1$ and $\langle z, h_2 \rangle = 1$, because $1 = \langle y, z \rangle = \langle z, y \rangle = \langle z, y \oplus z \rangle = \langle z, h_2 \rangle$. We get that $x \oplus h_1 \oplus h_2 = z \oplus h_2 = z \oplus (y \oplus z) = y$ hence $z = Z_{h_2} Z_{h_1} x$. Else, find indices $j, k \in 2, 4, \cdots, 2n$ where $(x_{j-1}, x_j) \neq (0,0)$ and $(y_{j-1}, y_j) = (0,0)$ and $(x_{k-1}, x_k) = (0,0)$ and $(y_{k-1}, y_k) \neq (0,0)$. Such indices exist because column vectors $x$ and $y$ are non zero. Find values for a pair $(z_{j-1}, z_j)$ such that $x_{j-1} z_j \oplus x_j z_{j-1} = 1$ and pair $(z_{k-1}, z_k)$ such that $y_{k-1} z_k \oplus y_k z_{k-1} = 1$, with all other elements of $z$ set to null. Again, we have $\langle x, z \rangle = 1$ and $\langle y, z \rangle = 1$. Furthermore, making $h_1 = x \oplus z$ and $h_2 = y \oplus z$, so are $\langle x, h_1 \rangle = 1$ and $\langle z, h_2 \rangle = 1$. We get that $x \oplus h_1 \oplus h_2 = z \oplus h_2 = z \oplus (y \oplus z) = y$ and $z = Z_{h_2} Z_{h_1} x$. $\quad\square$

**Symplectic Algorithm.** Firstly, let us define the set of symplectic pairs of column vectors as

$$S_i = \{(v, w) \in \mathbb{F}_2^{2i} \times \mathbb{F}_2^{2i} | <v, h> = 1\}.$$

Furthermore, let $n$ be a positive integer. We have the following sequence of subgroups $Sp(2) \subset Sp(4) \subset \cdots \subset Sp(2(n-1)) \subset Sp(2n)$, with $\sigma$ in $Sp(2(i-1))$ included in $Sp(2i)$ as the following constructed symplectic

$$\left( \begin{array}{c} \begin{bmatrix} 1 & 0 \\ 0 & 1 \end{bmatrix} \ 0 \cdots 0 \\ 0 \\ \vdots \quad \sigma \\ 0 \end{array} \right).$$

The *symplectic algorithm* [13] uses the equivalence between the set of symplectic pairs of column vectors $S_i$ and the set $Sp(2i)/Sp(2(i-1))$. In other words, a symplectic pair in $S_i$ represents a left coset in the set $Sp(2i)/Sp(2(i-1))$.

**Lemma 6.** *The cardinal of $S_i$ is $2^{2i-1} \cdot (4^i - 1)$.*

*Proof.* It follows from the Lagrange's theorem and ratio $|Sp(2i)|/|Sp(2(i-1))|$.

Given a symplectic pair $(v, w)$ in $S_1$, the corresponding symplectic in $Sp(2)$ is the matrix

$$\sigma_2 = \begin{pmatrix} v_1 & w_1 \\ v_2 & w_2 \end{pmatrix}.$$

The construction of symplectic matrices is inductive. For $i$ greater than one, given a pair $(v, w)$ in $S_i$ and a symplectic $\sigma_{2(i-1)}$ in $Sp(2i-1)$, the corresponding symplectic in $Sp(2i)$ is the matrix resulting from the composition

$$\sigma_{2i} = Z_s Z_{h_0} Z_{h_1} Z_{h_2} \left( \begin{array}{c} \begin{bmatrix} 1 & 0 \\ 0 & 1 \end{bmatrix} \ 0 \quad \cdots \quad 0 \\ 0 \\ \vdots \quad \sigma_{2(i-1)} \\ 0 \end{array} \right). \tag{6}$$

There are four transvections involved. Let $I_{*,j}$ denote the $j$-th column of the identity matrix of dimension $2i$. The transvections $Z_{h_1}$ and $Z_{h_2}$ are such that their application to $I_{*,1}$ is equal to $v$, i.e., $Z_{h_1} Z_{h_2} I_{*,j}$ is equal to $v$. The transvections $Z_{h_1}$ and $Z_{h_2}$ are calculated applying Lemma 5. Their application transvects the first column $I_{*,1}$ into $v$. By construction (detailed in the sequel), $Z_{h_0}$ has no effect on $v$. Because it is defined using $v$, the transvection $Z_s$ has no effect on $v$. Result, the first column of $\sigma_{2i}$ is $v$. The composition $Z_s Z_{h_0} Z_{h_1} Z_{h_2} I_{*,2}$ yields the column vector $w$. Result, the second column of $\sigma_{2i}$ is $w$. The transvections $Z_s$, $Z_{h_0}$, $Z_{h_1}$, and $Z_{h_2}$ are applied to the remaining $2(i-1)$ columns of the matrix in Eq. 6 embedding the matrix $\sigma_{2(i-1)}$.

The elements of $Sp(2i)$ are indexed in the range $0, \ldots, |Sp(2i)| - 1$. A symplectic index $k$ can be decomposed into two numbers

$$k = l \cdot |S_i| + m \text{ with } 0 \le l < |Sp(2(i-1))| \text{ and } 0 \le m < |S_i|. \tag{7}$$

The number $l$ is recursively used as the index of the symplectic $\sigma_{2(i-1)}$ in $Sp(2(i-1))$. The number $m$ is used to index the element of $S_i$, i.e., the symplectic pair determining the matrix $\sigma_{2i}$. The only issue that remains is mapping the index

$m$ to a symplectic pairs of the set $S_i$. Applying Lemma 6, the symplectic pair index $m$ can be decomposed into two numbers

$$m = r \cdot 2^{2i-1} + s \text{ with } 0 \le r < (2^{2i} - 1) \text{ and } 0 \le s < 2^{2i-1}. \tag{8}$$

The binary expansion of $r + 1$ over $2i$ bits becomes the column vector $v$. Let $b$ be the binary representation of $s$ over $2i - 1$ bits. It used to obtain the column vector $h_0$ in the transvection $Z_{h_0}$, that is,

$$h_0 = Z_{h_1} Z_{h_2} \left( I_{*,1} + b_2 I_{*,3} + \ldots b_{2i-1} I_{*,2i} \right).$$

The transvection $Z_s$ is defined with $s$ resulting from the composition $\neg b_1 v$. The vector $w$ is defined as

$$w = Z_s Z_{h_0} Z_{h_1} Z_{h_2} I_{*,2}.$$

Our MATLAB implementation of the Symplectic algorithm is available online here https://github.com/michelbarbeau/symplectic. Although, the detailed design is different, the logic is equivalent to the Python implementation of Koenig and Smolin [13]. The time complexity is still in $\mathcal{O}(n)$.

**Discussion on Adversary.** For quantum messages authenticated with the scheme described in Sect. 3, the adversary is challenged with finding a key in a domain in the order of $|\mathcal{C}(n)|$ plus $|G_n|$. To enlarge the search domain, a larger value for the security parameter $d$ can be chosen. For symmetric-key message encryption, Sect. 4, the adversary is challenged with finding an index in a domain in the order of $|G_{2n}|$. For asymmetric-key message encryption, Sect. 5, the adversary is challenged with finding a trapdoor-key in a domain in the order of $2^n$. All domains have exponential size cardinality, with respect to $n$.

**Lemma 7.** *The classical key size for quantum message authentication is*

$$n^2 + \sum_{i=1}^{n} \log_2(4^i - 1) + n \text{ bits}$$

*with the lower bound $2n^2$ bits and upper bound $2n(n+1)$ bits. For asymmetric-key and symmetric-key message encryption, the classical key size is $2n$ bits.*

*Proof.* For the upper and lower bounds, observe that $2(i-1) \le \log_2(4^i - 1) \le 2i$, for $i \ge 1$. Therefore

$$n(n-1) = 2 \sum_{i=1}^{n} (i-1) \le \sum_{i=1}^{n} \log_2(4^i - 1) \le 2 \sum_{i=1}^{n} i = n(n+1).$$

This lemma allows some comparison with the key size used in classical cryptography. However, the comparison has its limits. In contrast to classical cryptography, due to the fact that a state can be measured only once and is non-duplicative, on quantum data, the adversary has only one trial!

# 7    Conclusion

This paper has established connections between classical key material and quantum message authentication, asymmetric encryption and symmetric encryption. They use Clifford and Pauli quantum operators. Mapping of classical key material, i.e., classical bits, to quantum operators has been discussed. The key domain cardinals (key length) are exponential (polynomial) in size, with respect to the dimension of the quantum state. This work enables the use of classical key management for quantum data security.

**Acknowledgements.** I would like to thank my colleague Prof. Evangelos Kranakis for his advice on some aspects of this paper. This work was partially supported by the Natural Sciences and Engineering Research Council of Canada (NSERC).

# References

1. Aharonov, D., Ben-Or, M., Eban, E.: Interactive proofs for quantum computations. In: Proceedings of Innovations of Computer Science, pp. 453–469 (2010)
2. Alagic, G., Broadbent, A., Fefferman, B., Gagliardoni, T., Schaffner, C., Jules, M.S.: Computational security of quantum encryption. In: 9th International Conference on Information Theoretic Security (ICITS) (2016)
3. Bennett, C., Brassard, G., Crépeau, C., Jozsa, R., Peres, A., Wootters, W.: Teleporting an unknown quantum state via dual classical and einstein-podolsky-rosen channels. Phys. Rev. Lett. **70**, 1895–1899 (1993)
4. Bolt, B., Room, T.G., Wall, G.E.: On the Clifford collineation, transform and similarity groups. I. J. Aust. Math. Soc. **2**(1), 60–79 (1961)
5. Bolt, B., Room, T.G., Wall, G.E.: On the Clifford collineation, transform and similarity groups. II. J. Aust. Math. Soc. **2**(1), 80–96 (1961)
6. Briegel, H.J., Dür, W., Cirac, J.I., Zoller, P.: Quantum repeaters: the role of imperfect local operations in quantum communication. Phys. Rev. Lett. **81**, 5932–5935 (1998)
7. Broadbent, A., Wainewright, E.: Efficient simulation for quantum message authentication. In: 9th International Conference on Information Theoretic Security (ICITS) (2016)
8. Chevalley, C.: Theory of Lie Group. Dover Publications, Mineola (2018)
9. Diaconis, P., Shahshahani, M.: The subgroup algorithm for generating uniform random variables. Probab. Eng. Inf. Sci. **1**(1), 15–32 (1987)
10. Diffie, W., Hellman, M.: New directions in cryptography. IEEE Trans. Inf. Theory **22**(6), 644–654 (1976)
11. Gallian, J.: Contemporary Abstract Algebra, 8th edn. Cengage Learning - Brooks/Cole, Boston (2006)
12. Goldreich, O., Levin, L.A.: A hard-core predicate for all one-way functions. In: Proceedings of the Twenty-First Annual ACM Symposium on Theory of Computing, STOC 1989, pp. 25–32. ACM, New York (1989)
13. Koenig, R., Smolin, J.: How to efficiently select an arbitrary Clifford group element. J. Math. Phys. **55**, 122202 (2014)
14. Nielsen, M.A., Chuang, I.L.: Quantum Computation and Quantum Information. Cambridge University Press, Cambridge (2010)

15. St-Jules, M.: Secure quantum encryption. Master's thesis, School of Graduate Studies and Research, University of Ottawa, Ottawa, Ontario, Canada, November 2016
16. Wainewright, E.: Efficient simulation for quantum message authentication. Master's thesis, Faculty of Graduate and Postgraduate Studies, University of Ottawa, Ottawa, Ontario, Canada (2016)

# Continuous-Variable Quantum Network Coding Against Pollution Attacks

Tao Shang[1]([✉]), Ke Li[1], Ranyiliu Chen[2], and Jianwei Liu[1]

[1] School of Cyber Science and Technology, Beihang University, Beijing 100083, China
shangtao@buaa.edu.cn
[2] School of Electronic and Information Engineering, Beihang University,
Beijing 100083, China
yiliu_cr@buaa.edu.cn

**Abstract.** Continuous-variable quantum communication protocols have gained much attention for their ability to transmit more information with lower cost. To break through the bottleneck of quantum network coding schemes, continuous-variable quantum network coding (CVQNC) schemes were proposed. In spite of network throughput improvement, CVQNC also brings on security problems such as pollution attacks, in which case errors are accumulated and spread to all downstream nodes. In this paper, we propose a continuous-variable quantum network coding scheme with quantum homomorphic signature to resist pollution attacks. The scheme utilizes pre-shared quantum entanglement and classical communication to implement perfect crossing transmission of two quantum states. By combining two quantum signatures of classical messages generated by source nodes, the scheme will generate a homomorphic signature, which is used to verify the identities of different data sources in a quantum network. Security analysis shows the proposed scheme is secure against forgery and repudiation.

**Keywords:** Quantum network coding · Continuous variables · Pollution attacks · Quantum homomorphic signature

## 1 Introduction

Quantum information can be classified into two categories according to the spectrum characteristic of its eigenstate, namely discrete variables and continuous variables. Discrete variables, which have a discrete spectrum, denote quantum variables of finite-dimensional Hilbert space such as the polarization of single photons. Continuous variables, which have a continuous spectrum, denote quantum variables of infinite-dimensional Hilbert space such as the amplitude and phase quadratures of an optical field. From the perspective of practical application, quantum communication using continuous variables has prominent advantages over discrete variables. Concretely, the preparation, unitary manipulation and measurement of a continuous-variable quantum state can be efficiently implemented by basic quantum optics, while more sophisticated devices

© Springer Nature Switzerland AG 2019
S. Feld and C. Linnhoff-Popien (Eds.): QTOP 2019, LNCS 11413, pp. 196–206, 2019.
https://doi.org/10.1007/978-3-030-14082-3_17

are required in the discrete-variable case. Moreover, continuous-variable quantum communication is more efficient because continuous-variable quantum states can carry more information and are easier to generate and detect. Therefore, both theoretical and experimental investigations increasingly focus on continuous variables.

To break through the bottleneck of quantum networks, network coding technology [1] was introduced to quantum networks. XQQ (crossing two qubits) is the first quantum network coding (QNC) cheme, which indicates that quantum network coding is possible in the butterfly network and can realize crossing transmission of two arbitrary quantum states [2]. To date, many quantum network coding schemes have been proposed [3–5]. Most of them can be called discrete-variable quantum network coding (DVQNC) schemes, which use discrete variables as information carrier for QNC. To improve the transmission efficiency and practicability of quantum networks, continuous-variable quantum network coding (CVQNC) schemes were proposed [6]. With the help of pre-shared entanglement and classical communication, a CVQNC scheme can transmit quantum states with the fidelity of 1 and has great advantage over discrete-variable paradigms in network throughput from the viewpoint of classical information transmission.

Due to the encoding characteristics, CVQNC suffers from pollution attacks. If an error occurs, it will be spread to all downstream nodes because the error message is used to generate other messages by means of encoding. Consequently, target nodes cannot decode the correct message. In classical networks, homomorphic signature is utilized to resist pollution attacks, which allows an aggregator to generate a new signature by directly manipulating multiple original signatures. Recently, quantum homomorphic signature schemes have been proposed [7–9]. Among them, the scheme in the reference [9] is a continuous-variable quantum homomorphic signature (CVQHS) scheme. If CVQHS is feasible in CVQNC, it will be very helpful to improve the security of continuous-variable quantum network communication.

By introducing the CVQHS scheme [9] into the CVQNC scheme with prior entanglement [6], we design a secure quantum network coding scheme which can verify the identity of data sources and resist pollution attacks in the butterfly network.

## 2 Related Works

### 2.1 Continuous-Variable Quantum Network Coding

In 2017, Shang et al. [6] proposed the concept of CVQNC and designed two CVQNC schemes, namely a CVQNC scheme using approximate operations and a CVQNC scheme with prior entanglement. The CVQNC scheme using approximate operations has a low fidelity as quantum states cannot be cloned perfectly. With the help of pre-shared entanglement and classical communication, the CVQNC scheme with prior entanglement can transmit quantum states with a fidelity of 1. Obviously, the CVQNC scheme with prior entanglement is superior to the other scheme in practical use.

In the CVQNC scheme with prior entanglement scheme, two source nodes $s_1$ and $s_2$ share two EPR pairs. $s_1$ has the first mode of the EPR pairs, namely $|x_{11} + ip_{11}\rangle$ and $|x_{21} + ip_{21}\rangle$, and $s_2$ has the second mode $|x_{12} + ip_{12}\rangle$ and $|x_{22} + ip_{22}\rangle$. Here $x$ and $p$ denotes the continuous-variable basis conventionally. $s_1$ and $s_2$ prepare their coherent states $|x_A + ip_A\rangle$ and $|x_B + ip_B\rangle$. $t_1$ and $t_2$ are target nodes. It is described as follows:

**Step 1.** $s_1$ applies Bell detection and displacement. Mix $|x_A + ip_A\rangle$ and $|x_{21} + ip_{21}\rangle$ at a 50:50 beam splitter and then measure the $x$ quadrature and the $p$ quadrature of two output modes, respectively. Then $s_1$ displaces $|x_{11} + ip_{11}\rangle$ according to measurement results $(x_1, p_1)$. Measurement results are sent to $r_1$ via a classical channel and mode $|x'_{11} + ip'_{11}\rangle$ is sent to $t_2$ via a quantum channel.

Similarly, $s_2$ applies Bell detection and displacement, and sends measurement results $(x_2, p_2)$ to $r_1$ and the displaced mode $|x'_{22} + ip'_{22}\rangle$ to $t_1$.

**Step 2.** $r_1$ adds up the received classical numbers and sends the result $(x_1 + x_2) + i(p_1 + p_2)$ to $r_2$.

**Step 3.** $r_2$ copies the received classical message and sends replicas to $t_1$ and $t_2$.

**Step 4.** According to the received classical message, $t_k (k = 1, 2)$ displaces the quantum state $|\hat{x}'_{k\oplus 1, k\oplus 1} + i\hat{p}'_{k\oplus 1, k\oplus 1}\rangle$.

By setting squeezing parameter $r \to \infty$, $t_1$ and $t_2$ obtain $|x_A + ip_A\rangle$ and $|x_B + ip_B\rangle$, respectively.

## 2.2   Continuous-Variable Quantum Homomorphic Signature

Entanglement swapping [10] is a technique that realizes the entanglement of two quantum systems which have no direct interaction with each other. By operating on halves of two entangled states, the other halves are entangled. So it looks like entanglement is swapped among the particles. A continuous-variable approach for entanglement swapping has been proposed.

According to the idea that continuous-variable entanglement swapping ca be utilized to generate a homomorphic signature, Li et al. [9] proposed the first continuous-variable quantum homomorphic signature scheme. It generates quantum signatures for classical messages and generates homomorphic signatures from two original quantum signatures. So it can be applied to the butterfly network with quantum-classical channels.

The basic model is shown in Fig. 1. Assume there are two signers $A$ and $B$, an aggregator $M$ and a verifier $V$. The CVQHS scheme is defined by a tuple of algorithms (Setup, Sign, Combine, Verify) and is described as follows.

(1) Setup

    **Step 1.** $A$ shares two secret keys $k_{A_1}$ and $k_{A_2}$ with $V$ by continuous-variable quantum key distribution. Meanwhile, $B$ shares two secret keys $k_{B_1}$ and $k_{B_2}$ with $V$.

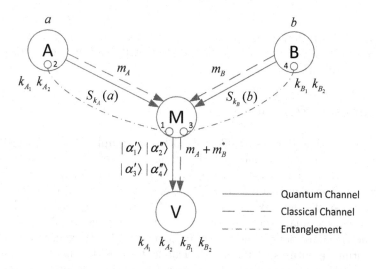

**Fig. 1.** Basic model of CVQHS

**Step 2.** $M$ prepares two pairs of entangled states, namely $(|\alpha_1\rangle, |\alpha_2\rangle)$ and $(|\alpha_3\rangle, |\alpha_4\rangle)$.

(2) Sign

**Step 1.** $A$ signs its classical message $a$ by displacing the quadratures of $|\alpha_2\rangle$, while $B$ signs its classical message $b$ by displacing the quadratures of $|\alpha_4\rangle$. The details of signing a continuous-variable quantum signature is presented in the reference [9].

**Step 2.** $A$ sends the signature $|\alpha_2'\rangle$ and the classical message $m_A$ to $M$, while $B$ sends the signature $|\alpha_4'\rangle$ and the classical message $m_B$ to $M$.

(3) Combine

**Step 1.** $M$ applies Bell detection on $|\alpha_1\rangle$ and $|\alpha_3\rangle$ and obtains the classical measurement results $x_1'$ and $p_3'$. At this point, $|\alpha_2'\rangle$ and $|\alpha_4'\rangle$ are entangled.

**Step 2.** $M$ mixes $|\alpha_2'\rangle$ and $|\alpha_4'\rangle$ at a 50:50 beam splitter and obtains two new signatures, denoted as $|\alpha''_2\rangle$ and $|\alpha''_4\rangle$.

**Step 3.** $M$ sends the quantum states $|\alpha_1'\rangle, |\alpha_2''\rangle, |\alpha_3'\rangle, |\alpha_4''\rangle$ and the classical message $m_{A+B} = m_A + m_B^*$ to $V$.

(4) Verify

**Step 1.** $V$ measures the $x$ quadrature of $|\alpha_2''\rangle$ and the $p$ quadrature of $|\alpha_4''\rangle$ by homodyne detection and obtains the measurement results $x''_2$ and $p''_4$.

**Step 2.** $V$ measures the $x$ quadrature of $|\alpha_1'\rangle$ and the $p$ quadrature of $|\alpha_3'\rangle$ by homodyne detection and obtains $x_1'$ and $p_3'$. Then $V$ calculates $x_V = \sqrt{2}(x_2'' - \tau x_1')$ and $p_V = \sqrt{2}(p_4'' - \tau p_3')$, where $\tau$ is the transmissivity of quantum channels.

**Step 3.** $V$ calculates $a$ and $b$ from the received classical message $m_{A+B} = m_A + m_B^*$. Then $V$ calculates $m_{k_{A_2}(B_2)}$, $x_V' = a + k_{A_1} + x_{k_{A_2}} + b + k_{B_1} + x_{k_{B_2}}$, and $p_V' = a + k_{A_1} + p_{k_{A_2}} - b - k_{B_1} - p_{k_{B_2}}$ according to pre-shared secret keys. To verify the authenticity and integrity of the signatures, $V$ calculates

$H_x = (x_V - \tau x'_V)^2$ and $H_p = (p_V - \tau p'_V)^2$. If $H_x \leq H_{th}$ and $H_p \leq H_{th}$, $V$ will confirm that $|\alpha''_2\rangle$ and $|\alpha''_4\rangle$ are the signatures of $M$ and accept the classical messages $a$ and $b$. Otherwise, $V$ will deny the signatures. $H_{th}$ is the verification threshold.

## 3   Proposed Scheme

Our scheme is based on the CVQNC scheme equipped with CVQHS. For CVQNC, a source node applies Bell detection to two of its quantum states and displaces another quantum state according to the measurement results. Then it sends the displaced quantum state to a target node and the measurement results to an intermediate node. With respect to CVQHS, in verifying phase the source node generates a quantum signature of the measurement results and sends it to the intermediate node. And the intermediate node will generate homomorphic quantum signatures. Before a target node decodes the quantum message, it must verify the received quantum signatures. The network setting is presented in Fig. 2. $s_1$ and $s_2$ are source nodes and signers, $r_1$ and $r_2$ are intermediate nodes, and $t_1$ and $t_2$ are target nodes and verifiers.

The proposed scheme is described as follows:

**Step 1.** Setup phase. $s_1$ shares secret keys $k_{A_1}$ and $k_{A_2}$ with target nodes. $s_2$ shares secret keys $k_{B_1}$ and $k_{B_2}$ with target nodes. $s_1$ and $s_2$ share two pairs of entangled states, namely $(|\alpha_{11}\rangle, |\alpha_{12}\rangle)$ and $(|\alpha_{21}\rangle, |\alpha_{22}\rangle)$, and $s_i(i = 1, 2)$ holds the $i$th modes of the entangled states. $r_1$ prepares two pairs of entangled states, namely $(|\alpha_1\rangle, |\alpha_2\rangle)$ and $(|\alpha_3\rangle, |\alpha_4\rangle)$. A pair of entangled states $(|\alpha_1\rangle, |\alpha_2\rangle)$ meet the following correlations:

$$\begin{cases} \hat{x}_1 = (e^r \hat{x}_{1(3)}^{(0)} + e^{-r} \hat{x}_2^{(0)})/\sqrt{2} \\ \hat{p}_1 = (e^{-r} \hat{p}_{1(3)}^{(0)} + e^r \hat{p}_2^{(0)})/\sqrt{2} \\ \hat{x}_2 = (e^r \hat{x}_{1(3)}^{(0)} - e^{-r} \hat{x}_2^{(0)})/\sqrt{2} \\ \hat{p}_2 = (e^{-r} \hat{p}_{1(3)}^{(0)} - e^r \hat{p}_2^{(0)})/\sqrt{2} \end{cases} \tag{1}$$

$$\begin{cases} \langle (\hat{x}_1 - \hat{x}_2)^2 \rangle = e^{-2r}/2 \\ \langle (\hat{p}_1 + \hat{p}_2)^2 \rangle = e^{-2r}/2 \end{cases} \tag{2}$$

where $\hat{x}_k^{(0)}$ and $\hat{p}_k^{(0)}$ ($k = 1, 2$) are a conjugate pair of quadratures of a vacuum state $|\alpha_k^{(0)}\rangle$ and $|\alpha_k^{(0)}\rangle = |x_k^{(0)} + ip_k^{(0)}\rangle$. Then $r_1$ sends $|\alpha_2\rangle$ to $s_1$ and $|\alpha_4\rangle$ to $s_2$.

**Step 2.** Encoding phase. $s_1$ applies Bell detection to $|\alpha_{21}\rangle$ and its signal mode $|\alpha_A\rangle$. Concretely, it mixes two modes at a 50:50 beam splitter and applies homodyne detection to the output states. Then it displaces the quadratures of $|\alpha_{11}\rangle$ according to the measurement results $(x_{A1}, p_{A1})$, where $x_{A1}$ is the measurement result of the $x$ quadrature of $|\alpha_A + \alpha_{21}\rangle$ and $p_{A1}$ the $p$ quadrature of $|\alpha_A - \alpha_{21}\rangle$. The displaced mode is denoted as $|\alpha'_{11}\rangle$.

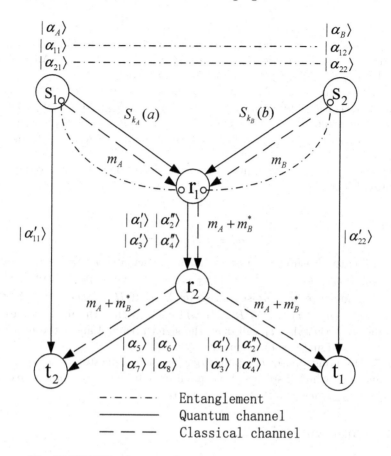

**Fig. 2.** CVQNC scheme with quantum homomorphic signature

Similarly, $s_2$ applies Bell detection to $|\alpha_{12}\rangle$ and its signal mode $|\alpha_B\rangle$. Then it displaces the quadratures of $|\alpha_{22}\rangle$ according to the measurement results $(x_{B2}, p_{B2})$. The displaced mode is denoted as $|\alpha'_{22}\rangle$.

**Step 3.** Signing phase. $s_1$ generates a real number $a$ from $(x_{A1}, p_{A1})$ according to an encoding rule which is predetermined among all nodes. Then $s_1$ uses secret keys $k_{A_1}$ and $k_{A_2}$ to generate a signature of $a$, which is denoted by $S_{k_A}(a)$. $S_{k_A}(a) = \hat{D}(m_a + m_{k_{A_1}} + m_{k_{A_2}})|\alpha\rangle$, where $\hat{D}(\gamma) = \exp(\gamma \hat{a}^\dagger - \gamma^* \hat{a})$ is the displacement operator. $m_a$, $m_{k_{A_1}}$ and $m_{k_{A_2}}$ are complex numbers, namely $m_a = a + ia$, $m_{k_{A_1}} = k_{A_1} + ik_{A_1}$, $m_{k_{A_2}} = x_{k_{A_2}} + ip_{k_{A_2}}$, where $x_{k_{A_2}}$ and $p_{k_{A_2}}$ satisfy

$$\begin{cases} x_{k_{A_2}} = k_{A_2}, p_{k_{A_2}} = 0 & \text{if } a + k_{A_2} \text{ is odd} \\ x_{k_{A_2}} = 0, p_{k_{A_2}} = k_{A_2} & \text{if } a + k_{A_2} \text{ is even} \end{cases} \tag{3}$$

$s_1$ sends $m_A = a + ia$ and $S_{k_A}(a)$ to $r_1$ and $|\alpha'_{11}\rangle$ to $t_2$.
Similarly, $s_2$ generates a real number $b$ from $(x_{B2}, p_{B2})$ and generates its signature $S_{k_B}(b)$. Then $s_2$ sends $m_B = b + ib$ and $S_{k_B}(b)$ to $r_1$ and $|\alpha'_{22}\rangle$ to $t_1$.

**Step 4.** Combining phase. $r_1$ applies Bell detection to $|\alpha_1\rangle$ and $|\alpha_3\rangle$, and denotes output states as $|\alpha'_1\rangle$ and $|\alpha'_3\rangle$. Then $r_1$ mixes $S_{k_A}(a)$ and $S_{k_A}(a)$ at a 50:50 beam splitter and denotes output states as $|\alpha''_2\rangle$ and $|\alpha''_4\rangle$. After that, $r_1$ sends $|\alpha'_1\rangle$, $|\alpha'_3\rangle$, $|\alpha''_2\rangle$, $|\alpha''_4\rangle$, and $m_A + m_B^*$ to $r_2$.

**Step 5.** Copying phase. After applying homodyne detection to the received quantum states, $r_2$ prepares quantum states $|\alpha_5\rangle$, $|\alpha_6\rangle$, $|\alpha_7\rangle$, and $|\alpha_8\rangle$ according to measurement results, where $x_5 = x'_1$, $p_6 = p'_3$, $x_7 = x''_2$, and $p_8 = p''_4$. Then $r_2$ sends $|\alpha'_1\rangle$, $|\alpha'_3\rangle$, $|\alpha''_2\rangle$, and $|\alpha''_4\rangle$ to $t_1$, and sends $|\alpha_5\rangle$, $|\alpha_6\rangle$, $|\alpha_7\rangle$, and $|\alpha_8\rangle$ to $t_2$.

**Step 6.** Verifying phase. $t_1$ applies homodyne detection to the received quantum states and calculates

$$x_V = \sqrt{2}(x''_2 - \tau x'_1)$$

and

$$p_V = \sqrt{2}(p''_4 - \tau p'_3)$$

according to measurement results. Then $t_1$ calculates $x'_V = a + k_{A_1} + x_{k_{A_2}} + b + k_{B_1} + x_{k_{B_2}}$ and $p'_V = a + k_{A_1} + p_{k_{A_2}} - b - k_{B_1} - p_{k_{B_2}}$ according to the classical message and pre-shared secret keys. If $H_x = (x_V - \tau x'_V)^2 \leq H_{th}$ and $H_p = (p_V - \tau p'_V)^2 \leq H_{th}$, $t_1$ will confirm that the messages are from $s_1$ and $s_2$. Otherwise, $t_1$ will deny the signatures and abort the protocol. $t_2$ verifies the signatures in a similar way.

**Step 7.** Decoding phase. $t_1$ applies displacement operator to $|\alpha'_{22}\rangle$ so as to obtain $|\alpha_A\rangle$, which displaces the $x$ quadrature by $x_{A1} + x_{B2}$ and the $p$ quadrature by $p_{A1} + p_{B2}$. Similarly, $t_2$ obtains $|\alpha_B\rangle$.

## 4   Scheme Analysis

### 4.1   Performance Analysis

In this section, we will analyze the performance of the proposed scheme from the perspectives of fidelity and network throughput.

**Fidelity.** Here we consider the quantum state at the target node $t_1$. The case of the target node $t_2$ will be the same for the reason of symmetry. Assume the entangled states shared between two source nodes are ideal, i.e., perfectly correlated and maximally entangled, $r \to \infty$.

After Step 2, the two quadratures of $|x'_{22} + ip'_{22}\rangle$ are

$$\begin{cases} \hat{x}'_{22} = \hat{x}_{22} - \hat{x}_B + \hat{x}_{12} \\ \hat{p}'_{22} = \hat{p}_{22} - \hat{p}_B - \hat{p}_{12} \end{cases}. \tag{4}$$

At $t_1$, $\hat{x}'_{22}$ is displaced as

$$\begin{aligned} \hat{x}'_{22} \to \hat{x}''_{22} &= \hat{x}'_{22} + \sqrt{2}(\hat{x}_{A1} + \hat{x}_{B2}) \\ &= \hat{x}_A - \hat{x}_{21} + \hat{x}_{22} \\ &= \hat{x}_A - \sqrt{2}e^{-r}\hat{x}_2^{(0)} \end{aligned}. \tag{5}$$

Similarly, $\hat{p}'_{22}$ is displaced as

$$\begin{aligned}
\hat{p}'_{22} \to \hat{p}''_{22} &= \hat{p}'_{22} + \sqrt{2}(\hat{p}_{A1} + \hat{p}_{B2}) \\
&= \hat{p}_A + \hat{p}_{21} + \hat{p}_{22} \\
&= \hat{p}_A + \sqrt{2}e^{-r}\hat{p}_1^{(0)}
\end{aligned} \tag{6}$$

When $r$ increases to infinity, the final quantum state at $t_1$, namely $|\hat{x}''_{22}+i\hat{p}''_{22}\rangle$, becomes $|\hat{x}_A + i\hat{p}_A\rangle$, which is the same as the quantum state sent by $s_1$. As a result, we can conclude that our CVQNC scheme can successfully transmit two quantum states across perfectly by a single network use. The fidelity of the scheme is 1.

**Network Throughput.** Assume that a coherent state $|x + ip\rangle$ is modulated with classical characters, i.e., $x, p \in \{0, 1, \ldots, N - 1\}$. When the classical character set for modulation has $N$ elements, each character contains $\log_2 N$ bits of information. In the proposed CVQNC scheme, each target node receives one coherent state with a fidelity of 1. So each target node can receive $2\log_2 N$ bits of classical information by a single network when applying the proposed CVQNC scheme.

As a matter of fact, coherent states are nonorthogonal, which means they cannot be perfectly distinguished to yield the ideal entropy calculated. The square of the inner product of two arbitrary coherent states $|\alpha\rangle$ and $|\beta\rangle$ is

$$|\langle \beta | \alpha \rangle|^2 = e^{-|\alpha-\beta|^2}. \tag{7}$$

Equation (7) shows that coherent states $|\alpha\rangle$ and $|\beta\rangle$ are approximately orthogonal when $|\alpha - \beta| \gg 1$, so they can be measured by heterodyne detection with high accuracy. The condition $|\alpha - \beta| \gg 1$ requires the elements of classical character set to have large values, which may be impractical for implementation.

## 4.2   Security Analysis

Our CVQNC scheme utilizes quantum homomorphic signature to resist pollution attacks. To successfully tamper or forge messages, an attacker Eve or dishonest intermediate nodes must forge a signature that can pass verification. Meanwhile, a legitimate source node may attempt to deny that it has sent a message to a target node. In this section, we will analyze the security of the proposed scheme from the perspectives of unforgeability and non-repudiation.

**Unforgeability.** Firstly, we analyze whether secret keys can be calculated on the basis of the classical messages and quantum states transmitted in the channels.

By eavesdropping the link $s_1 \to r_1$, an attacker Eve cannot calculate the secret keys $k_{A_1}$ and $k_{A_2}$ on the basis of $m_A = a+ia$ and $|\alpha_2+m_A+m_{k_{A_1}}+m_{k_{A_2}}\rangle$. Similarly, for the link $B \to r_1$, the secret keys $k_{B_1}$ and $k_{B_2}$ cannot be calculated.

By eavesdropping the link $r_1 \to r_2 \to t_1$, Eve can intercept the classical message $m_A + m_B^*$ and the quantum states

$$\begin{cases} |\frac{1}{\sqrt{2}}\alpha_+\rangle \\ |\frac{1}{\sqrt{2}}(\alpha_+^* + m_A + m_B + m_{k_{A_1}} + m_{k_{A_2}} + m_{k_{B_1}} + m_{k_{B_2}})\rangle \\ |\frac{1}{\sqrt{2}}\alpha_-\rangle \\ |\frac{1}{\sqrt{2}}(\alpha_-^* + m_A - m_B + m_{k_{A_1}} + m_{k_{A_2}} - m_{k_{B_1}} - m_{k_{B_2}})\rangle \end{cases} . \tag{8}$$

Owing to the famous uncertainty principle, two quadratures $x$ and $p$ cannot be precisely measured at the same time. To calculate as much information as possible, Eve needs to measure the $x$ quadrature of part of the quantum states and the $p$ quadrature of the other part of the quantum states. Without loss of generality, we assume Eve measures the $x$ quadrature of $|\alpha_1'\rangle$ and $|\alpha_2''\rangle$ and the $p$ quadrature of $|\alpha_3'\rangle$ and $|\alpha_4''\rangle$. Since Eve does not have any copy of the quantum states, it can only calculate $k_{A_1} + x_{k_{A_2}} + k_{B_1} + x_{k_{B_2}}$ and $k_{A_1} + p_{k_{A_2}} - k_{B_1} - p_{k_{B_2}}$ on the basis of the measurement results and $m_A + m_B^*$. So the secret keys $k_{A_1}$, $k_{A_2}$, $k_{B_1}$, and $k_{B_2}$ cannot be calculated.

Secondly, we analyze whether an attacker Eve or a dishonest intermediate node $r_2$ can forge the signature of a legitimate source node.

In the verifying phase, $t_1$ and $t_2$ use pre-shared secret keys to verify a signature. So Eve and $r_2$ must obtain secret keys to forge a signature that can pass verification. It has been proved that Eve and $r_2$ cannot calculate secret keys on the basis of the classical messages and the quantum states transmitted in the channels. Assume the secret keys are distributed securely in the setup phase, then it is impossible for Eve and $r_2$ to have the secret keys. So Eve and $r_2$ cannot forge the signature of a legitimate signer.

In fact, even if Eve and $r_2$ obtain the secret keys in the setup phase, they cannot forge the signature of a legitimate signer because they do not share entangled states with $r_1$. Assume Eve or $r_2$ has a quantum state $|\alpha_0\rangle = |x_0 + ip_0\rangle$ and the secret keys of $A$, namely $k_{A_1}$ and $k_{A_2}$. It signs a message $e$ with secret keys $k_{A_1}$ and $k_{A_2}$ and generates the signature $S_{k_A}^E(e) = |\alpha_0 + m_E + m_{k_{A_1}} + m_{k_{A_2}}\rangle = |\alpha_E'\rangle$, where $m_E = e + ie$. Then it substitutes the classical message and the signature of $A$ with $m_E$ and $|\alpha_E'\rangle$, respectively. In the verifying phase, $t_1$ calculates

$$\begin{cases} x_V = x_0 - x_1 + e + k_{A_1} + x_{k_{A_2}} + b + k_{B_1} + x_{k_{B_2}} \\ p_V = p_1 - p_0 + e + k_{A_1} + p_{k_{A_2}} - b - k_{B_1} - p_{k_{B_2}} \end{cases} \tag{9}$$

and

$$\begin{cases} x_V' = e + k_{A_1} + x_{k_{A_2}} + b + k_{B_1} + x_{k_{B_2}} \\ p_V' = e + k_{A_1} + p_{k_{A_2}} - b - k_{B_1} - p_{k_{B_2}} \end{cases} . \tag{10}$$

It is obvious that $x_V \neq x_V'$ and $p_V \neq p_V'$. $t_1$ confirms the existence of an attacker or a dishonest intermediate node and denies the signatures. The case of $t_2$ will be the same for the reason of symmetry. In conclusion, Eve and $t_2$ cannot forge the signature of a legitimate source node.

Thirdly, we analyze whether a dishonest intermediate node $r_1$ can forge the signatures of a legitimate source node under the assumption of secure secret key distribution.

According the assumption that secret keys are distributed securely, $r_1$ cannot obtain the secret keys $k_{A_1}$, $k_{A_2}$, $k_{B_1}$ and $k_{B_2}$. Instead, $r_1$ can only calculate $k_{A_1} + x_{k_{A_2}} + k_{B_1} + x_{k_{B_2}}$ and $k_{A_1} + p_{k_{A_2}} - k_{B_1} - p_{k_{B_2}}$.

Assume $r_1$ substitutes $m_{A+B}$ with a fake message $m_{A+B}^M$. It needs to prepare two quantum states $|\alpha_2^M\rangle = |\frac{1}{\sqrt{2}}(\alpha_+^* + m_2^M)\rangle$ and $|\alpha_4^M\rangle = |\frac{1}{\sqrt{2}}(\alpha_-^* + m_4^M)\rangle$ to substitute the original signatures $|\alpha_2''\rangle$ and $|\alpha_4''\rangle$, respectively. Here, $m_2^M$ and $m_4^M$ are complex numbers and expressed as $m_2^M = x_2^M + ip_2^M$ and $m_4^M = x_4^M + ip_4^M$. In the verifying phase, $t_1$ measures quantum states and calculates $x_V = x_2^M$ and $p_V = p_4^M$. According to $m_{A+B}^M$, $t_1$ calculates $a'$ and $b'$ that satisfy $a' + b' + i(a' - b') = m_{A+B}^M$. Then $m'_{k_{A_{1(2)}}} = x'_{k_{A_{1(2)}}} + ip'_{k_{A_{1(2)}}}$ can be calculated according to the pre-shared secret keys $k_{A_2}$ and $k_{B_2}$. After that, $t_1$ calculates $x_V' = x_{A+B}^M + k_{A_1} + x'_{k_{A_2}} + k_{B_1} + x'_{k_{B_2}}$ and $p_V' = p_{A+B}^M + k_{A_1} + p'_{k_{A_2}} - k_{B_1} - p'_{k_{B_2}}$. Finally, $t_1$ calculates $H_x = (x_V - \tau x_V')^2$ and $H_p = (p_V - \tau p_V')^2$. If $H_x \leq H_{th}$ and $H_p \leq H_{th}$, $t_1$ accepts the signatures. Otherwise, $t_1$ denies the signatures.

To make the fake signatures pass verification, $r_1$ should choose $m_2^M$, $m_4^M$, and $m_{A+B}^M$ to satisfy

$$\begin{cases} x_2^M = x_{A+B}^M + k_{A_1} + x'_{k_{A_2}} + k_{B_1} + x'_{k_{B_2}} \\ p_4^M = p_{A+B}^M + k_{A_1} + p'_{k_{A_2}} - k_{B_1} - p'_{k_{B_2}} \end{cases}. \tag{11}$$

Since $r_1$ cannot obtain $k_{A_2}$ and $k_{B_2}$, it cannot calculate the correct values for $m'_{k_{A_1}}$ and $m'_{k_{A_2}}$. So $r_1$ cannot forge the signatures of a legitimate source node.

**Non-repudiation.** Assume secret keys are distributed securely in the setup phase and the target nodes are honest. It has been proved that an attacker Eve and dishonest intermediate nodes cannot perform forgery, so only the signatures generated by pre-shared secret keys can pass verification. It has also been proved that secret keys cannot be calculated, so nobody but legitimate source nodes and target nodes can obtain the secret keys. Since the target nodes are honest, they always announces the correct verification results and will not forge signatures. Therefore, a source node cannot repudiate its signature after it has passed verification.

## 5    Conclusion

In this paper, we proposed a continuous-variable quantum network coding scheme against pollution attacks. The scheme is based on the CVQNC scheme with prior entanglement and can transmit two quantum states across perfectly. By combining continuous-variable quantum homomorphic signature, our scheme can verify the identity of different data sources. As long as quantum signatures pass verification, target nodes can decode their quantum states and obtain the correct messages. Security analysis shows that our scheme is secure against forgery and repudiation. The proposed CVQNC scheme against pollution attacks is a basic model for constructing a secure quantum network and future work is needed to explore higher security.

**Acknowledgment.** This project was supported by the National Natural Science Foundation of China (No. 61571024) for valuable helps.

# References

1. Ahlswede, R., Cai, N., Li, S.R., Yeung, R.W.: Network information flow. IEEE Trans. Inf. Theory **46**(4), 1204–1216 (2000)
2. Hayashi, M., Iwama, K., Nishimura, H., Raymond, R., Yamashita, S.: Quantum network coding. In: Thomas, W., Weil, P. (eds.) STACS 2007. LNCS, vol. 4393, pp. 610–621. Springer, Heidelberg (2007). https://doi.org/10.1007/978-3-540-70918-3_52
3. Hayashi, M.: Prior entanglement between senders enables perfect quantum network coding with modification. Phys. Rev. A **76**(4), 1–5 (2007)
4. Leung, D., Oppenheim, J., Winter, A.: Quantum network communication-the butterfly and beyond. IEEE Trans. Inf. Theory **56**, 3478–3490 (2010)
5. Satoh, T., Gall, F.L., Imai, H.: Quantum network coding for quantum repeaters. Phys. Rev. A **86**(3), 1–8 (2012)
6. Shang, T., Li, K., Liu, J.W.: Continuous-variable quantum network coding for coherent states. Quantum Inf. Process. **16**(4), 107 (2017)
7. Shang, T., Zhao, X.J., Wang, C., Liu, J.W.: Quantum homomorphic signature. Quantum Inf. Process. **14**(1), 393–410 (2015)
8. Luo, Q.B., Yang, G.W., She, K., Li, X.Y., Fang, J.B.: Quantum homomorphic signature based on Bell-state measurement. Quantum Inf. Process. **15**(12), 5051–5061 (2016)
9. Li, K., Shang, T., Liu, J.W.: Continuous-variable quantum homomorphic signature. Quantum Inf. Process. **16**(10), 246 (2017)
10. Zukowski, M., et al.: Event-ready-detectors Bell experiment via entanglement swapping. Phys. Rev. Lett. **71**(26), 4287 (1993)

# On the Influence of Initial Qubit Placement During NISQ Circuit Compilation

Alexandru Paler[✉]

Linz Institute of Technology, Johannes Kepler University, Linz, Austria
alexandru.paler@jku.at

**Abstract.** Noisy Intermediate-Scale Quantum (NISQ) machines are not fault-tolerant, operate few qubits (currently, less than hundred), but are capable of executing interesting computations. Above the quantum supremacy threshold (approx. 60 qubits), NISQ machines are expected to be more powerful than existing classical computers. One of the most stringent problems is that computations (expressed as quantum circuits) have to be adapted (compiled) to the NISQ hardware, because the hardware does not support arbitrary interactions between the qubits. This procedure introduces additional gates (e.g. SWAP gates) into the circuits while leaving the implemented computations unchanged. Each additional gate increases the failure rate of the adapted (compiled) circuits, because the hardware and the circuits are not fault-tolerant. It is reasonable to expect that the placement influences the number of additionally introduced gates. Therefore, a combinatorial problem arises: how are circuit qubits allocated (placed) initially to the hardware qubits? The novelty of this work relies on the methodology used to investigate the influence of the initial placement. To this end, we introduce a novel heuristic and cost model to estimate the number of gates necessary to adapt a circuit to a given NISQ architecture. We implement the heuristic (source code available on github) and benchmark it using a standard compiler (e.g. from IBM Qiskit) treated as a black box. Preliminary results indicate that cost reductions of up to 10% can be achieved for practical circuit instances on realistic NISQ architectures only by placing qubits differently than default (trivial placement).

## 1 Introduction

Quantum computing has become a practical reality, and the current generation of machines includes a few tens of qubits and supports only non-error corrected quantum computations. These machines, called noisy intermediate-scale quantum (NISQ), are not operating perfectly, meaning that each executed quantum gate introduces a certain error into the computed results. NISQ machines have started being used for investigations which would have been computationally very difficult previously (e.g. [1]). However, such investigations/computations are

© Springer Nature Switzerland AG 2019
S. Feld and C. Linnhoff-Popien (Eds.): QTOP 2019, LNCS 11413, pp. 207–217, 2019.
https://doi.org/10.1007/978-3-030-14082-3_18

not fault-tolerant. It is not possible to mitigate the error through error detecting and correcting codes, because too many qubits would be required.

Executing quantum computations has become streamlined such that quantum computers are placed in the cloud and used through dedicated software tools (e.g. Qiskit and Cirq). These tools implement almost an entire work flow – from describing quantum computations in a higher level language to executing them in a form compiled for specific architectures. Computed results are influenced by the *number and type of gates* from the executed low level circuit – output values become indistinguishable from random noise if the compiled circuits are too deep.

It is complex to compile NISQ circuits which have a minimum number of gates – this is a classical design automation problem. Recent work on the preparation (compilation) of NISQ circuits (e.g. [2]) treat this problem as a search problem. Therefore, following questions are meaningful: (1) which search criteria has to be optimised?; (2) where should the search start from?; (3) does the starting point influence the quality of the compiled circuits? Answering these questions is based on some assumptions highlighted in the following sections.

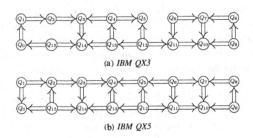

(a) *IBM QX3*

(b) *IBM QX5*

**Fig. 1.** Two architectures of IBM quantum machines where qubits are drawn as vertices and supported CNOT gates as arrows between qubit pairs: (a) QX3 cannot interact through CNOT qubit pairs (5,6) and (2,15); (b) QX5 has a grid like structure. In both diagrams the arrow indicate the direction of the supported CNOT: the target qubit is indicated by the tip of the arrow. The graph representation of the architectures is called *coupling graph/map*.

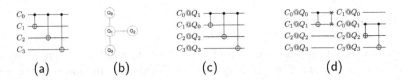

(a)          (b)          (c)          (d)

**Fig. 2.** Quantum circuit examples: (a) a four qubit (wire) quantum circuit; (b) a four qubit coupling graph. The initial placement of wires on hardware qubits influences the number of inserted SWAP gates: (c) No SWAPS inserted; (d) One SWAP inserted.

## 2   Background

The reader is assumed to be familiar with the elements of quantum information processing and quantum circuits (wires, gates, qubits etc.). Generally, there are no restrictions how to place gates in a quantum circuit. For example, any pair of circuit wires may be used for applying a CNOT. However, NISQ machines are not flexible, and support only the application of two-qubit gates (CNOTS, in the following) between certain pairs of hardware qubits. In Fig. 1, hardware qubits are drawn as vertices and the arrows represent CNOTs. Only some qubit pairs can be used for executing CNOTs (cf. QX3 with the QX5 architecture): CNOT between qubit pairs (2,15) and (5,6) are not supported by QX3. QX5 is also restrictive: for example, it does not allow a CNOT between the qubit pair (0,13). Currently available architectures do not allow *bidirectional* CNOTs (any qubit from the pair can be either the control or the target), but it is expected that this issue will not be of concern in the future. The introduced heuristic and cost model assume undirected *coupling graphs* (e.g. Fig. 2b).

*NISQ circuit compilation* refers to the preparation of a quantum circuit to be executed on a NISQ machine expressed as a coupling graph. This procedure consists of two steps: (1) choosing an initial placement of circuit qubits to hardware qubits; (2) ensuring that the gates from the circuit can be executed on the machine. The terms machine, hardware and computer will be used interchangeably.

The *qubit placement problem* is to choose which circuit wire is associated to which hardware qubit. This problem is also known as the *qubit allocation* problem. For a quantum circuit with $n$ qubits (wires) to be executed on a NISQ machine with $n$ hardware qubits (vertices in the coupling graph), let the circuit qubits be $C_i$ and the machine qubits $Q_i$ with $i \in [0, n-1]$.

A *trivial placement* is to map $C_0$ on $Q_0, C_1$ on $Q_1$ etc. One can express the placement as a vector $pl = [0, 1 \ldots, n-1]$ where wire $C_i$ is placed on hardware qubit $Q_i = pl[C_i]$. Vector $pl$ is a permutation of $n$ elements, and there are $n!$ ways how the circuit can be initially mapped to the machine. The search method (mentioned in Sect. 1) need not use a too exact cost heuristic, because of the following assumption (a speculation).

**Assumption 1.** *There are many initial placements, from all the $n!$ possibilities, which generate optimal compiled circuits.*

After the initial placement is determined, each gate from the input circuit is mapped to the hardware. Until recently the ordering of the gates has been assumed fixed, but recent works (e.g. [3]) investigate how gate commutativity rules influence the compilation results. This work does not make any assumptions about how gates are mapped to the hardware. The placement of the circuit qubits dictates where SWAP gates are introduced.

*Example 1.* Consider the circuit from Fig. 2a that has to be executed on the architecture from Fig. 2b. If wire $C_0$ is placed on $Q_0$, the CNOT between $C_0$ and $C_2$ cannot be executed before a SWAP is inserted (Fig. 2d).

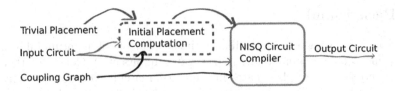

**Fig. 3.** Methodology: the compiler (orange) has three inputs and an output (the compiled circuit). One of the inputs is the initial placement of qubits (dotted green box). The potential influence of the initial placement is investigated by switching on and off the proposed heuristic. (Color figure online)

### 2.1 Motivation

Qubit placement has not been considered relevant for compilation optimisation, but the significance of this problem will increase when the hardware includes many qubits. Although NISQ machines and their associated circuits are small, it is not computationally feasible to compile circuits using exact methods. This makes it difficult to determine the exact cost of an optimal solution, because finding out the exact cost means that a potential solution has already been computed. Compilation strategies will rely on heuristics and search algorithms [4] (e.g. hill climbing, A*). As a result, the cost of compiling a circuit to NISQ should be evaluated as fast as possible and should be used as a search guiding heuristic during compilation.

### 2.2 Previous Work

Cost models for the effort of adapting a circuit to a given architecture have been intensively researched in the community (e.g. [5]), but few focused on non-exact methods and cost models (e.g. not using constraint satisfaction engines, or Boolean formula satisfiability). Most of the previous works considered regular architectures (1D or 2D nearest neighbour).

Previous works (e.g. [2] or [6]) did not specifically address the initial placement problem. Nevertheless, one of the latest works where the importance of the problem is mentioned but not tackled is [7]. Consequently, this work is, to the best of our knowledge, the first to specifically propose a solution to this problem, and to analyse its relevance in a quantitative and compiler independent manner. Independently to this work, the article of [8] was published at the same time with our paper. The difference is that we propose a different, simpler heuristic, and our methodology enables a fair interpretation of the placement influence, because the compiler is treated as a black box (Fig. 3).

## 3   Methods

This work investigates the relevance of a heuristic to approximate the number of necessary SWAPs. The three questions from Sect. 1 are reformulated into a single one: **Is it possible to choose the initial placement of circuit wires on hardware qubits, such that the compiler (without being modified) generates more optimal quantum circuits?**

To answer this question we propose: (a) a heuristic for the estimation of the costs, and (b) a greedy search algorithm to chose the best initial placement. The output of our procedure (dotted green box in Fig. 3) is a qubit placement to be used by an unmodified NISQ circuit compiler. The default Qiskit swap_mapper compiler was chosen, because it is a randomised algorithm. The compiler is treated as a black box with three inputs and an output (Fig. 3).

In order to simplify the presentation, the heuristic and the search method are broken into concrete steps presented in the following subsections. The proposed methodology uses all the information available: (1) input circuit $circ$, (2) coupling graph $g$, and (3) a given placement vector (a permutation) $pl$. In the following only CNOTs $cn_i$ from $circ$ are considered, and $i$ is their position index in the gate list of $circ$.

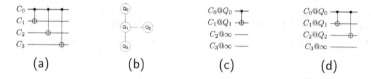

**Fig. 4.** Methodology examples: (a) a four qubit (wire) quantum circuit; (b) a four qubit coupling graph. Sub-circuits are resulting by considering partial placements: (c) $[Q_0, Q_1, \infty, \infty]$ results in a circuit with a single active CNOT; (d) $[Q_0, Q_1, Q_2, \infty]$ results in a sub-circuit with two active CNOTs.

### 3.1 Number of Active CNOTs

**Assumption 2.** *It is possible to extract a sub-circuit by removing wires and all the gates associated with the removed wires. Such a sub-circuit does not represent the original computation, but can be used to estimate costs. The minimum sub-circuit has only two wires.*

For a circuit of $n$ wires, a sub-circuit is obtained using a *partial placement*, where not placed wires are signaled by $\infty$. For the example of a circuit with $n = 4$ wires to be mapped on an architecture with four hardware qubits, $[Q_0, \infty, \infty, Q_1]$ means that $C_0$ is placed on $Q_0$, $C_3$ on $Q_1$, while $C_1$ and $C_2$ are not placed.

For a given partial placement $pl$, the *number of active CNOTs $nr(pl)$* is the number of CNOTs operating on placed wires. In other words, this number is an indication of the number of gates $cn$ obtained from the initial circuit after removing all the wires marked by $\infty$ in $pl$.

$$nr(circ, pl) = |\{cn_i \text{ appl. to } C_q \text{ s.t. } pl[C_q] \neq \infty\}| \qquad (1)$$

*Example 2.* The sub-circuits from Fig. 4c and d are obtained from the initial circuit from Fig. 4a by considering partial placements.

## 3.2   Tracking the Distance

The distances between hardware qubits is computed by a Floyd Warshall algorithm applied to $g$. For a given CNOT $cn_i$ and a coupling graph $g$, a distance function $dist(c_i, pl, g)$ computes the distance between the hardware qubits determined from $pl$. The computation method is presented instead of an expression for $dist$. The $dist$ function employs an $offset$ register of the same dimension as the number of hardware qubits.

**Assumption 3.** *For a $cn_i$ having its wires placed on $Q_c$ and $Q_t$, the distance between $Q_c$ and $Q_t$ is $d_{c,t}$ which includes the Floyd Warshall distance and the values for $Q_c$ and $Q_t$ from $offset$. Both $Q_c$ and $Q_t$ have to be moved (swapped) $m = d_{c,t}/2$ times, where $m > 0$. The $offset$ of $min(Q_c, Q_t)$ is updated to increase by $m - 1$, and the $offset$ of $max(Q_c, Q_t)$ decreases by $m$.*

The placement of the CNOT wires introduces a number of SWAP gates equal to the distance minus one. The approximation is introduced by stating that, if wire $C_w$ is placed on $pl[C_w] = Q_h$, after a sequence of CNOTs, where $Q_h$ generated $offset[Q_h]$ SWAPS, the actual hardware qubit of $C_w$ is $Q_h + offset[Q_h]$. This approximation is not ensured to be pessimistic [4].

## 3.3   Attenuation

CNOTs $cn_i$ earlier in the gate list (low values of $i$) of $circ$ should be preferred for activation instead of later CNOTs (higher values of $i$). A similar mechanism was very recently independently proposed in [8]. Attenuation is a prototypical mechanism to control the sensitivity of the cost function and a linear function was chosen. However, other functions may be more appropriate – future work.

$$att(cn_i, circ) = \frac{|circ| - i}{|circ|} \tag{2}$$

## 3.4   Approximate Cost Function

For an input circuit $circ$, a (partial) placement $pl$, and a coupling graph $g$, the approximate cost in terms of SWAPs needed to compile it to $g$ is given by:

$$err(circ, pl, g) = \sum_{i=0}^{|circ|} att(cn_i, circ) \times dist(cn_i, pl, g) \tag{3}$$

$$cost(circ, pl, g) = \frac{err(circ, pl, g)}{nr(circ, pl)} \tag{4}$$

As a result, for equal $err$ values, the cost is lower for partial placements that activate more CNOTs from the initial circuit. Simultaneously, for equal number of active CNOTs, lower total errors are preferred.

## 3.5   Greedy Search - One Hardware Qubit After the Other

The search procedure for finding an initial placement is iterative: it places a circuit wire on each $Q_i$ at a time. The search starts with $Q_0$, and a best wire candidate for this hardware qubit. For simplicity, assume the wire is $C_0$. Afterwards, the search starts to construct partial placements that minimise the $cost(circ, pl, g)$ function. For example, assume the best wire from $circ$ is $C_4$, because it increases the number of activated CNOTs in the sub-circuit generated by the partial placement $pl$ while keeping the error low. Thus, $C_4$ is placed on $Q_1$ (hardware qubit index following 0 is 1) s.t. $pl[C_4] = Q_1$.

The implemented search method generates a search tree. The root is the wire placed on $Q_0$, and the leaves are the wires placed on $Q_n$. Choosing the initial placement with the minimal cost is similar to executing a breadth first search, and this could lead to an exponential complexity. Nevertheless, the search algorithm was designed to limit the number of children (maximum number of candidates with equal estimated costs) and supports also cut-off heuristics (chose the best partial placement, when the search tree has a certain number of levels). The search heuristic is implemented modularly (can be entirely replaced) in our Python implementation.

## 3.6   Discussion: Heuristic vs. Exact Solution

It can be argued that the herein presented cost model and heuristic to estimate the costs can be replaced by exact methods (e.g. SAT solvers). This is true for circuits with few qubits, and this could allow an improved quantitative analysis of the proposed methodology.

However, we argue that small circuits do not require design automation methods, because these can be designed manually (an approach taken by Cirq[1]). NISQ platforms are very resource restricted and it is faster to design circuits by hand, than to encode all the constraints into a SAT instance. According to Assumption 1, it is also expected (based on previous experience in designing quantum circuits) that exact methods will be slower and error prone (due to software bugs), such that software verification tools are necessary – or manual verification of the compiled circuits is required, which again is an argument against exact methods.

For experimentation using cloud based services (e.g. IBM[2]) it is important to compile as well as possible, but *compilation speed plays an important role*, too. Therefore, exact methods are not scalable (fast enough) and their practicality is restricted to quantitative comparisons on a set of benchmark circuits. At this point it should be noted that the research community has not agreed on a set of NISQ benchmarking circuits. The circuits we used in this paper were chosen only because they have a practical relation to existing IBM architectures.

---

[1] github.com/quantumlib/Cirq.

[2] quantumexperience.ng.bluemix.net/qx.

It is impractical to manually design large NISQ circuits (e.g. more than 100 qubits), and at the same time it is not possible (due to scalability limitations) to use exact methods for such circuits.

Consequently, we argue that exact solutions are theoretically useful, but impractical. Heuristics are the only practical and scalable solution for the compilation and optimisation of NISQ circuits. Therefore, *heuristics can be compared only against heuristics*. This work, besides [8], is one of the very few to tackle qubit placement from a heuristics point of view.

## 4    Results

The proposed heuristic for choosing the initial placement was implemented in Python and integrated into the Qiskit SDK challenge framework[3]. Simulation results were obtained for four circuits: two random circuits (rand0_n16_d16 and rand1_n16_d16) of 16 qubits, and two practically relevant circuits (qft_n16 and qubits_n16_exc). Each of the four circuits was compiled for both the IBM QX3 and QX5 architectures. Each compilation was performed twice: the first time with the trivial initial placement, and the second time with a placement chosen by the presented heuristic. Two variants of the heuristics were used: one with attenuation and the second without. A total of 250 simulations were performed for each combination of circuit, architecture, and placement. Each combination used the same seed – for example, simulation number 15 used a seed value of 15.

**Table 1.** Simulation results with attenuation (Sect. 3.3). The $T$ column is the trivial placement cost, and $H$ the initial placement computed by the heuristic. The *Med.* columns include the median value from 250 simulations, and the *Avg.* columns the average value. *Imp.* is computed as the ratio between *T.Avg.* and *H.Avg.*

| Circuit | Arch. | T.Avg. | T.Med. | H.Avg. | H.Med. | Imp. |
|---|---|---|---|---|---|---|
| qft_n16 | QX3 | 7837.76 | 7811.00 | 7469.70 | 7387.00 | 1.05 |
| qft_n16 | QX5 | 7142.37 | 7092.00 | 7265.49 | 7178.00 | 0.98 |
| qubits_n16_exc | QX3 | 1447.22 | 1422.00 | 1541.15 | 1522.00 | 0.94 |
| qubits_n16_exc | QX5 | 562.00 | 562.00 | 562.00 | 562.00 | 1.00 |
| rand0_n16_d16 | QX3 | 13094.18 | 13078.00 | 13492.68 | 13420.00 | 0.97 |
| rand0_n16_d16 | QX5 | 12370.19 | 12326.00 | 12243.39 | 12261.00 | 1.01 |
| rand1_n16_d16 | QX3 | 14188.84 | 14209.00 | 13604.65 | 13569.00 | 1.04 |
| rand1_n16_d16 | QX5 | 12779.25 | 12810.00 | 12542.53 | 12537.00 | 1.02 |

The results validate to some degree the predictive power of the approximative cost model (see previous Section), and indicates the sensitivity of future compilation search algorithms to the initial placement of the circuit qubits.

---

[3] https://qe-awards.mybluemix.net/static/challenge_24_1_2018.zip.

**Table 2.** Simulation results without attenuation. The $T$ column is the trivial placement cost, and $H$ the initial placement computed by the heuristic. The *Med.* columns include the median value from 250 simulations, and the *Avg.* columns the average value. *Imp.* is computed as the ratio between $T.Avg.$ and $H.Avg.$

| Circuit | Arch. | T.Avg. | T.Med. | H.Avg. | H.Med. | Imp. |
|---|---|---|---|---|---|---|
| qft_n16 | QX3 | 7837.76 | 7811.00 | 7205.67 | 7137.00 | 1.09 |
| qft_n16 | QX5 | 7142.37 | 7092.00 | 6859.87 | 6867.00 | 1.04 |
| qubits_n16_exc | QX3 | 1447.22 | 1422.00 | 1763.40 | 1773.00 | 0.82 |
| qubits_n16_exc | QX5 | 562.00 | 562.00 | 1050.51 | 1020.00 | 0.53 |
| rand0_n16_d16 | QX3 | 13094.18 | 13078.00 | 13013.54 | 12995.00 | 1.01 |
| rand0_n16_d16 | QX5 | 12370.19 | 12326.00 | 12085.82 | 12080.00 | 1.02 |
| rand1_n16_d16 | QX3 | 14188.84 | 14209.00 | 12851.42 | 12879.00 | 1.10 |
| rand1_n16_d16 | QX5 | 12779.25 | 12810.00 | 12121.74 | 12175.00 | 1.05 |

The costs from the Tables 1 and 2 are the sum of all specific gate type costs of each output circuit gate. The CNOTs have a cost of 10, single qubit gates a cost of 1 (the exception is formed by single qubit Z-axis rotations which have cost zero). The SWAP gates are decomposed into three CNOTs and four Hadamard gates. The columns with $T$ refer to the trivial placement, and $H$ to the initial placement computed by the heuristic. The *Med.* columns include the median value from the 250 simulations, and the *Avg.* columns the average value. The *Imp.* columns are computed as the ratio between the values from the corresponding $T.Avg.$ and $H.Avg..$ For example, an improvement of 1.05 means that the heuristic reduced cost by 5% (positive influence), while 0.55 means that the costs increased by 45% (negative influence).

Across all experiments, the median and the average values are almost equal, suggesting that there are no outliers. Thus, the observed improvements (positive and negative) seem correlated with the usage of the heuristic. Some preliminary conclusions are: (1) for practical circuits, the heuristic without attenuation generates diametrically opposed results (10% and 4% improvements for QFT and $-18\%$ and $-47\%$ increases for the other circuit) – this is *a good result in the case of QFT, because it is a widely used sub-routine in quantum algorithms*; (2) *the proposed heuristic seems to work better for architectures which do not have a completely regular structure* (cf. QX3 to QX5) – this is a good feature, because such architectures are the complex situations; (3) on average, the heuristic does not influence the overall performance of the compiler, but *for specific circuit types it has significant influence* (while some are improved by up to 10%, others are worsened by 45%) – the proposed heuristic is sensitive to a structural property of the circuits which needs to be determined.

The qubits_n16_exc circuit is a kind of black sheep of the experimented circuits: the proposed algorithm does not perform better than the defaults. Preliminary investigations show that this is happening because the circuit has a very regular structure, it actually uses only 14 qubits from 16 (leaving two un-operated, although defined in the quantum registers of the circuit), and the ordering of the qubits is (by chance?) already optimal for the considered architectures. Thus, incidentally, the default layout is already optimal and all the heuristic efforts are worsening the qubit allocation. Future work will look at different architectures.

## 5    Conclusion

The results are promising when comparing them with the ones obtained by the winner of the IBM Qiskit competition[4]: a specialised compiler using advanced techniques achieved on average 20% improvements. The proposed initial place-ment method generates improvements between 2% and 10% with a sub-optimal compiler, which IBM intended to replace by organising a competition. Our source code is available at github.com/alexandrupaler/k7m.

We did not perform a comparison against the IBM challenge winner, because it would not have been a fair one. The challenge winner uses information about the internals of the circuits (two qubit arbitrary gates are decomposed with the KAK algorithm [9]) in order to partially recompile subcircuits. The herein proposed cost model and search method are agnostic to quantum circuit internals or other previous circuit decomposition techniques.

Future work will focus on comparing future heuristics that will be proposed in the literature. We will first investigate why the structure of the qubits_n16_exc circuit makes the heuristic perform in a suboptimal manner (cf. discussion about benchmark circuits in Sect. 3.6). Second, different attenuation factor formulas will be investigated and compared to the very recent work of [8]. Finally, and most important, the effect of the heuristic will be analysed when used from within a compiler – do such approximate cost models improve the speed and the quality of NISQ circuit compilation? The current preliminary results presented in this work suggest this could be the case. It may be necessary to improve the distance estimation and the attenuation formulas.

**Acknowledgment.** The author thanks Ali Javadi Abhari for suggesting some of the circuits, and Lucian Mircea Sasu for very helpful discussions. This work was supported by project CHARON funded by Linz Institute of Technology.

## References

1. Viyuela, O., Rivas, A., Gasparinetti, S., Wallraff, A., Filipp, S., Martin-Delgado, M.A.: Observation of topological Uhlmann phases with superconducting qubits. npj Quantum Inf. 4(1), 10 (2018)
2. Zulehner, A., Paler, A., Wille, R.: Efficient mapping of quantum circuits to the IBM QX architectures. In: Design, Automation & Test in Europe Conference & Exhibition (DATE), 2018. IEEE (2018)

---

[4] www.ibm.com/blogs/research/2018/08/winners-qiskit-developer-challenge.

3. Hattori, W., Yamashita, S.: Quantum circuit optimization by changing the gate order for 2D nearest neighbor architectures. In: Kari, J., Ulidowski, I. (eds.) RC 2018. LNCS, vol. 11106, pp. 228–243. Springer, Cham (2018). https://doi.org/10.1007/978-3-319-99498-7_16
4. Russell , S., Norvig, P.: Artificial Intelligence: A Modern Approach (2002)
5. Saeedi, M., Markov, I.L.: Synthesis and optimization of reversible circuits-a survey. ACM Comput. Surv. (CSUR) **45**(2), 21 (2013)
6. Venturelli, D., Do, M., Rieffel, E., Frank, J.: Compiling quantum circuits to realistic hardware architectures using temporal planners. Quantum Sci. Technol. **3**(2), 025004 (2018)
7. Paler, A., Zulehner, A., Wille, R.: NISQ circuit compilers: search space structure and heuristics, arXiv preprint arXiv:1806.07241 (2018)
8. Li, G., Ding, Y., Xie, Y.: Tackling the qubit mapping problem for nisq-era quantum devices, arXiv preprint arXiv:1809.02573 (2018)
9. Zulehner, A., Wille, R.: Compiling SU (4) quantum circuits to IBM QX architectures, arXiv preprint arXiv:1808.05661 (2018)

# Towards a Pattern Language
# for Quantum Algorithms

Frank Leymann$^{(\boxtimes)}$ (iD)

IAAS, University of Stuttgart, Universitätsstr. 38, 70569 Stuttgart, Germany
`Frank.Leymann@iaas.uni-stuttgart.de`

**Abstract.** Creating quantum algorithms is a difficult task, especially for computer scientist not used to quantum computing. But quantum algorithms often use similar elements. Thus, these elements provide proven solutions to recurring problems, i.e. a pattern language. Sketching such a language is a step towards establishing a software engineering discipline of quantum algorithms.

**Keywords:** Quantum algorithms · Pattern languages ·
Software engineering

## 1 Introduction

### 1.1 Patterns and Pattern Languages

There is a significant difference in how quantum algorithms are presented and invented, and the way how traditional algorithms are built. Thus, computer scientists and software developers used to solve classical problems need a lot of assistance when being assigned to build quantum algorithms. To support and guide people in creating solutions in various domains, pattern languages are established. A *pattern* is a structured document containing an abstract description of a proven solution of a recurring problem. Furthermore, a pattern points to other patterns that may jointly contribute to an encompassing solution of a complex problem. This way, a network of related patterns, i.e. a *pattern language*, results. This notion of pattern and pattern language has its origin in [1]. Although invented to support architects in building houses and planning cities, it has been accepted in several other domains like pedagogy, manufacturing, and especially in software architecture (e.g. [13]).

In this paper, we lay the foundation for a pattern language for quantum algorithms. The need for documenting solutions for recurring problems in this domain can be observed in text books like [19, 21] that contain unsystematic explanations of basic "tricks" used in quantum algorithms. Our contribution is to systematize this to become a subject of a software engineering discipline for quantum algorithms.

© Springer Nature Switzerland AG 2019
S. Feld and C. Linnhoff-Popien (Eds.): QTOP 2019, LNCS 11413, pp. 218–230, 2019.
https://doi.org/10.1007/978-3-030-14082-3_19

## 1.2   Structure of a Pattern Document

A document specifying a pattern within a certain domain follows a fixed structure (e.g. [1,13]). While this structure may vary from domain to domain, many elements are in common, i.e. independent of the domain: Each pattern has a *name*. This name should be descriptive, identifying the problem to be solved.

The *intend* of the pattern briefly describes the goal to be achieved with the solution described by the pattern.

An *icon* represents the pattern visually, e.g. as a mnemonic. While the gate model makes heavy use of icons (as quantum gates and their wiring), other models don't; pattern icons may add a visual aspect to such models. Even for the gate model, patterns often abstract gates and their compositions, i.e. pattern icons are a more abstract representation of parts of an algorithm.

The *problem* statement concisely summarizes the problem solved by the pattern. Based on this, the reader can immediately decide whether the pattern is relevant for the problem at hand to be solved.

The *context* describes the situation or the forces, respectively, that led to the problem. It may refer to other patterns already applied.

The *solution* is the most important element of the pattern: it specifies in an abstract manner how to solve the problem summarized in the problem statement. The problem statement together with the solution is the underpinning of the pattern. Variants of solutions may be described depending on different flavors of the context.

The *known uses* section refers to algorithms that make use of the pattern. It confirms that the problem is recurring and that the presented solution is proven.

Other patterns related to the current one are referenced in the next element. These references link the individual patterns into a pattern language.

## 1.3   Overview

The patterns we propose in Sect. 2 are derived from algorithms based on the gate model. This is not a restriction in principle, because patterns based on other models may be added in future. Furthermore, patterns based on one model may be transformed into equivalent patterns of other models (e.g. the gate model is known to be equivalent to the measurement-based model [16]). Finally, we briefly indicate in Sect. 3 how patterns may be used in developing quantum algorithms.

## 2   Patterns for Quantum Algorithms

In this section we describe an initial set of basic patterns and their relations. Note, that this pattern language is far from being encompassing, i.e. it is expected that this pattern language evolves over time.

## 2.1    Initialization (aka State Preparation)

**Intend:** At the beginning of a quantum algorithm, the quantum register manipulated by the algorithm must be initialized. The initialization must be as easy as possible, considering requirements of the steps of the algorithm.

 How can the input of a quantum register be initialized in a straight-forward manner, considering immediate requirements of the following steps of the quantum algorithm?

**Context:** An algorithm typically requires input representing the parameters of the problem to be solved. Most quantum algorithms encode this input as part of the unitary transformations making up the quantum algorithm. E.g. if the overall algorithm is $U = U_n \circ \ldots \circ U_i \circ U_{i-1} \circ \ldots \circ U_1$, then $U_1, \ldots, U_{i-1}$ are operators that furnish the register to hold the parameters of the problem solved by the following operators $U_i, \ldots, U_n$. However, the initial state operated on by $U_{i-1} \circ \ldots \circ U_1$ must be set; $U_{i-1} \circ \ldots \circ U_1$ is called *state preparation* [25].

**Solution:** Often, the register will be initialized as the unit vector $|0 \ldots 0\rangle$. This register may have certain ancilla bits or workspace bits distinguished that are used to store intermediate results, to control the processing of the algorithm etc.

For example, the register is initialized with $|0\rangle^{\otimes n}|0\rangle^{\otimes m}$ (where the second part of the register consists of workspace bits) in order to compute the function table of the Boolean function $f : \{0,1\}^n \to \{0,1\}^m$.

An initialization with $|0\rangle^{\otimes n}|1\rangle$ supports to reveal membership in a set which is defined based on an indicator function (used to solve decision problems, for example) by changing the sign of the qbits representing members of this set.

Based on these simple initializations, more advanced states can be prepared. For example, [7] discusses several algorithms to load classical bits into a quantum register. [25] presents how to load a complex vector, [8] how to load a real vector based on corresponding data structures; thus, a matrix can be loaded as a set of vectors [18].

**Known Uses:** All algorithms must be initialized somehow.

**Next:** Often, after initialization the register must be brought into a state of *uniform superposition*. *Function tables* require the initializations discussed here. An initialized register may become input to an *oracle*.

## 2.2    Uniform Superposition

**Intend:** Typically, the individual qbits of a quantum register have to be in multiple states at the same time without preferring any at these states at the beginning of the computation.

 How can an equally weighted superposition of all possible states of the qbits of a quantum register be created?

**Context:** One origin of the power of quantum algorithms stems from quantum parallelism, i.e. the ability of a quantum register to represent multiple values at the same time. This is achieved by bringing (a subset of) the qbits of a quantum register into superposition. Many algorithms assume that at the beginning this superposition is uniform, i.e. the probability of measuring any of the qbits is the same.

**Solution:** Uniform superposition is achieved by initializing the quantum register as the unit vector $|0 \dots 0\rangle$ and applying the Hadamard transformation afterwards:

$$H^{\otimes n} \left( |0\rangle^{\otimes n} \right) = \frac{1}{\sqrt{2^n}} \sum_{x=0}^{2^n - 1} |x\rangle$$

In case the quantum register includes ancilla bits or workspace bits in addition to the computational basis, the computational basis is brought into superposition as described. The other bits may be brought into superposition themselves or not. This is achieved by using a tensor product $H^{\otimes n} \otimes U$, where $H^{\otimes n}$ operates on the computational basis and $U$ operates on the other bits (e.g., $U = I$ in case the other bits are not brought into superposition).

**Known Uses:** Most algorithms make use of uniform superposition.

**Next:** Creating uniform superposition makes use of *initialization*. A register in uniform superposition may be *entangled*. A register in uniform superposition may be input to an *oracle*.

## 2.3   Creating Entanglement

**Intend:** A strong correlation between qbits of a quantum register is often needed in order to enable algorithms that offer a speedup compared to classical algorithms.

 How can an entangled state be created?

**Context:** Entanglement is one of the causes of the power of quantum algorithms (see [5], although entanglement is not a necessity [3]). A quantum algorithm showing exponential speedup requires entanglement [17]. Thus, after initialization of a quantum register it should often be entangled for its further processing.

**Solution:** Several approaches can be taken to create an entangled state. For example, assume a binary function $f : \{0,1\}^n \rightarrow \{0,1\}^m$ and the corresponding unitary operation

$$U_f : \{0,1\}^{n+m} \rightarrow \{0,1\}^{n+m}, U_f \left( |x, y\rangle \right) = |x, y \oplus f(x)\rangle.$$

Then the following state is entangled:

$$U_f\left(H^{\otimes n} \otimes I^{\otimes m}\right)\left(|0\rangle \otimes |0\rangle\right)$$

With $f = id$ it is $U_{id} = CNOT$, which shows that $CNOT((H \otimes I)(|0\rangle \otimes |0\rangle))$ is entangled.

**Known Uses:** Many algorithms make use of entanglement.

**Next:** Typically, *initialization* precedes the creation of entanglement. A *function table* results from the above creation of entanglement based on $U_f$.

## 2.4   Function Table

**Intend:** Some problems can be reduced to determining global properties of a function. For that purpose, the corresponding function table should be computed efficiently and made available for further analysis.

> How can a function table of a finite Boolean function be computed?

**Context:** In order to compute the function table of a function $f : \{0,1\}^n \to \{0,1\}^m$, a classical algorithm requires to invoke the function for each value of the domain. Quantum parallelism allows to compute the values of such a finite Boolean function as a whole in a single step. This can be used to speedup finding global properties of the corresponding function. Note, that in case $m = 1$ the Boolean function is often an indicator function used to determine solutions of a decision problem.

**Solution:** The quantum register is split into the computational basis (the domain of the function f) consisting of n qbits x, and a workspace consisting of m qbits y, which is used to hold the values of f. Based on this, the unitary operator

$$U_f|x,y\rangle = |x, y \oplus f(x)\rangle$$

is defined.

After initializing the register with $|0\rangle^{\otimes n}|0\rangle^{\otimes m}$, the computational basis is brought into uniform superposition via $H^{\otimes n}$ leaving the workspace unchanged, and then the operator $U_f$ is applied only once resulting in the function table:

$$|0\rangle^{\otimes n}|0\rangle^{\otimes m} \xrightarrow{H^{\otimes n} \otimes I} \left(\frac{1}{\sqrt{2^n}} \sum_x |x\rangle\right) \otimes |0\rangle^{\otimes m} \xrightarrow{U_f} \frac{1}{\sqrt{2^n}} \sum_x |x\rangle|f(x)\rangle.$$

In case of an indicator function f (e.g. if f is representing a decision problem), the register is initialized with $|0\rangle^{\otimes n}|1\rangle$. Uniform superposition of the complete register is furnished by $H^{\otimes n+1}$. Applying $U_f$ finally results in

$$|0\rangle^{\otimes n}|1\rangle \xrightarrow{H^{\otimes n} \otimes H} \left( \frac{1}{\sqrt{2^n}} \sum_x |x\rangle \right) \otimes |-\rangle \xrightarrow{U_f} \left( \frac{1}{\sqrt{2^n}} \sum_{x=0}^{2^n-1} (-1)^{f(x)}|x\rangle \right) \otimes |-\rangle.$$

Thus, members of the computational basis indicate by their sign whether they are detected by the indicator function (minus sign) or not (plus sign) - aka "phase kickback".

**Known Uses:** The algorithms of Deutsch, Deutsch-Jozsa, Grover, Shor and others make use of function tables.

**Next:** Function tables require *initialization* discussed before. *Uniform superposition* of the computational basis is established before the function table is computed. *Amplitude amplification* is a generalization of function tables. The computation is performed by an *oracle*. Often, *uncompute* is required to continue processing.

### 2.5   Oracle (aka Black Box)

**Intend:** Quantum algorithms often need to compute values of a function f without having to know the details how such values are computed.

 How can the computation of another quantum algorithm be reused?

**Context:** Divide-and-Conquer is a well-established method in computer science to simplify the solution of complex problems. The concept of an oracle (or black box) as a granule of reuse with hidden internals supports this method for building quantum algorithms.

**Solution:** Oracles are used in problem specific manners. [14] discusses various kinds of oracles. Limitations of using oracles are discussed in [26].

**Known Uses:** The algorithms of Deutsch, Deutsch-Jozsa, Bernstein-Vazirani, Simon, Grover and others make use of oracles. See [20] for further usages. Next: An oracle often requires to uncompute its result state, and assumes a properly prepared register as input (initialization).

### 2.6   Uncompute (aka Unentangling aka Copy-Uncompute)

**Intend:** Often, entanglement of the computational basis of a quantum register with temporary qbits (ancilla, workspace) has to be removed to allow proper continuation of an algorithm.

How can entanglement be removed that resulted from a computation?

**Context:** A computation often needs temporary qbits, and at the end of the computation these qbits are entangled with the computational basis. This hinders access to the actual result of the computation, especially if the computation was used just as an intermediate step within an algorithm.

For example, if the computation should produce $\sum \alpha_i |\phi_i\rangle$ but in fact it produces $\sum \alpha_i |\phi_i\rangle |\psi_i\rangle$, the temporary qbits $|\psi_i\rangle$ can not be simply eliminated unless

$$\sum \alpha_i |\phi_i\rangle |\psi_i\rangle = \left( \sum \alpha_i |\phi_i\rangle \right) \otimes |\psi_i\rangle,$$

i.e. unless the computational basis and the temporary qbits are separable.

**Solution:** Most algorithms map $|x\rangle |0\rangle |0\rangle$ to $|x\rangle |g(x)\rangle |f(x)\rangle$ to compute a function f [8]. I.e. the second qbits represent a workspace that contains garbage g(x) at the end of the computation. This garbage has to be set to $|0\rangle$ to allow for proper continuation, especially if future parts of the algorithm expects the workspace to be initialized again by $|0\rangle$. More precisely, assume the computation $U_f$ resulted in

$$|x\rangle |0\rangle |0\rangle \underset{U_f}{\longrightarrow} \sum \alpha_y |x\rangle |y\rangle |f(x)\rangle,$$

i.e. the garbage state is $|g(x)\rangle = \sum \alpha_y |y\rangle$. Now, a fourth register initialized to $|0\rangle$ is added, and CNOT is applied (bitwise) to this fourth register controlled by the third register: this copies f(x) to the fourth register and $\sum \alpha_y |x\rangle |y\rangle |f(x)\rangle |f(x)\rangle$ results. Next, $U_f^{-1}$ is applied to the first three registers, giving $|x\rangle |0\rangle |0\rangle |f(x)\rangle$. Then, SWAP is applied to the last two registers leaving $|x\rangle |0\rangle |f(x)\rangle |0\rangle$. This now allows to discard the last register leaving $|x\rangle |0\rangle |f(x)\rangle$ as wanted (more details in [8]). [23] discusses how to use uncompute in several situations.

**Known Uses:** Deutsch-Joza, the HHL algorithm [15], quantum walks, realizations of classical circuits as quantum algorithms etc. make use of uncompute.

**Next:** An *oracle* often produces a state that is an entanglement between the computational basis and some temporary qbits, thus requires uncompute. A *function table* may be seen as a special case of an oracle.

## 2.7   Phase Shift

**Intend:** In a given register certain qbits should be emphasized.

 How can important aspects of a state been efficiently distinguished?

**Context:** When an algorithm is applied iteratively, and each iteration is assumed to improve the solution, those parts of the solution that did improve should be indicated. A phase shift can be such an indication.

**Solution:** The following operator $S_G^\phi$ can be efficiently implemented (see [24]) in terms of number of gates used:

$$\sum_{x=0}^{N-1} \alpha_x |x\rangle \, S_G^\phi \xrightarrow{} \sum_{x \in G} e^{i\phi} \alpha_x |x\rangle + \sum_{x \notin G} \alpha_x |x\rangle.$$

This operator shifts the qbits in $G \subseteq \{0, \ldots, N-1\}$ (the qbits improved: "good set") by phase $\phi$ and leaves the other qbits unchanged. There is even a variant of the operator that shifts the phases of the qbits in the good set by different values, i.e. $\phi = \phi(x)$.

**Known Uses:** The algorithms of Grover, Deutsch-Jozsa etc. use a phase shift.

**Next:** A *function table* based on an indicator function is a phase shift, with G as the set of base vectors qualifying under the indicator function. An *amplitude amplification* makes use of two phase shifts. A phase shift is used as an *oracle*.

## 2.8 Amplitude Amplification

**Intend:** Based on an approximate solution, the probability to find the precise solution should be increased from run to run of an algorithm U.

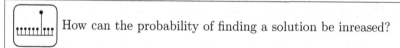 How can the probability of finding a solution be inreased?

**Context:** The function table of an indicator function f may list all solutions of a problem (i.e. $f(x) = 1 \Leftrightarrow x$ solves the problem). By measuring the corresponding state, a solution is found with a certain probability. But measuring destroys the state, i.e. if a solution is not received by measurement, the computation has to be performed again to support another new measurement.

Thus, a mechanism is wanted that doesn't need measurements and that allows to continue with the state achieved in case a solution is not found.

**Solution:** State is transformed in such a way that values of interest get a modified amplitude such that they get a higher probability of being measured after a couple of iterations [4].

The phase shift $S_G^\pi$ changes the sign of the phase of elements in G, the phase shift $S_0^\pi$ changes the sign of (the start value of the iteration) and leaves the other elements unchanged. Let U be the algorithm for computing approximate solutions (not using any measurements). Define the following as unitary operation:

$$Q = -U S_0^\pi U^{-1} S_G^\pi$$

If U is an algorithm that succeeds with a solution with probability t, $1/t$ iterations are required on the average to find a solution. $U|0\rangle$ is assumed to have a non-zero amplitude in G, otherwise no speedup can be achieved. If U has this property, Q will produce a solution within $O(\sqrt{1/t})$ iterations - which is a quadratic speedup. The number of iterations to be performed with Q is about

$$\frac{\pi}{4}\frac{1}{P_G U|0\rangle},$$

where $P_G$ is the projection onto the subspace spanned by G.

**Known Uses:** The algorithms of Grover and Simons, for example, make use of amplitude amplification. Also, the HHL algorithm for solving linear equations [15] uses this pattern. The state preparation algorithm of [25] uses amplitude amplification too. [4] discusses more algorithms making use of it.

**Next:** Part of the unitary operation Q is the function table $S_G^\pi$, which is also a special case of a *phase shift*. Amplitude amplifications are used as *oracles*.

## 2.9    Speedup via Verifying

**Intend:** Verifying whether a claimed solution is correct or not is sometimes simple. Such verifications may then be used to speedup solving a corresponding problem.

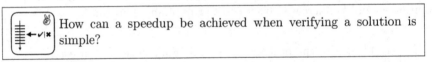 How can a speedup be achieved when verifying a solution is simple?

**Context:** Often, it is hard to find a solution of a problem, but verifying whether a claimed solution is correct is simple. For example, factorizing a number is hard, but multiplying numbers is simple. Thus, when a given list of prime numbers is claimed to be the factorization of a certain number, multiplying the prime numbers and comparing the result with the certain number is a simple way of verification.

**Solution:** Solving certain problems can be speedup by first listing all possible solutions, then scanning through the list and verifying whether the current member of the list at hand is a solution or not.

The verification of the possible solutions is done via an oracle. Scanning is done by means of the Grover algorithm, thus, $O(\sqrt{N})$ invocations of the Oracle function determines the solution. A prerequisite of this pattern is that solutions can be detected by means of an oracle.

**Known Uses:** Cracking keys, finding Hamiltonian cycles, solving 3-SAT, the Traveling Salesman Problem etc. can be approached this way.

**Next:** The verification is performed as an oracle.

## 2.10   Quantum-Classic Split

**Intend:** The solution of a problem is often not achieved by only using a quantum computer. Thus, the solution is performed partially on a classical computer and partially on a quantum computer, and both parts of the solution interact.

 How can a solution be split between a quantum computer and a classical computer?

**Context:** Some quantum algorithms inherently require pre- or post-processing on a classical device, resulting in a split of the solution into a classical part and a quantum part.

Also, if a quantum computer has a low number of qubits or its gates are noisy, a solution of a problem may have to be separated into a part executed on a quantum computer and a part executed on a classical computer [22].

**Solution:** The sheer fact that a split of the algorithms may be done is important. How such a split is applied is problem dependent.

**Known Uses:** Shor's algorithm or Simon's algorithm inherently make use of classical post-processing. The algorithm in [11] to solve combinatorial optimization problems uses classical pre-processing. The algorithm of [2] uses a split into a quantum part of the solution and a classical part to enable factorization on NISQ devices.

**Next:** Data is passed from the classical part of the solution to the quantum part by proper *initialization*.

# 3   Using Patterns

## 3.1   Patterns in Software Engineering

In software engineering, pattern languages exist in a plethora of domains like object orientation, enterprise integration, cloud computing etc. Typically, these pattern languages are delivered as books or on web pages. Software engineers determine their problem to solve and find a corresponding entry point into the pattern language. This first entry pattern links to other patterns that might be helpful in the problem context, if applicable these patterns are inspected and used too, their links are followed etc. This way, a subgraph (aka "solution path" [27]) of the pattern language is determined that conceptually solves a complex problem. Next, the abstract solutions of the patterns of this subgraph have to be implemented (i.e. turned into concrete solutions) so that they can be executed in a computing environment.

To make this process more efficient, a pattern repository can be used [12]: in essence, a pattern repository is a specialized database that stores pattern documents and manages the links between them. It allows to query the database to find appropriate patterns (e.g. to determine the entry pattern corresponding to a problem), supports browsing the content of each pattern document, and enables navigating between patterns based on the links between them.

In practice, patterns of several domains are needed to solve a complex problem [9]. For example, building an application for a cloud environment based on microservices that must fulfill certain security requirements leans on the corresponding three pattern languages (for cloud, microservices, security). For this purpose, patterns of a pattern language of a certain domain may point to patterns of another domain.

Similarly, the pattern language for quantum algorithm can be represented in a pattern repository. The corresponding patterns may be linked with patterns from concrete quantum programming languages to support the programming of the patterns.

### 3.2 Abstract Solutions and Concrete Solutions

Patterns describe abstract solutions, independent of any concrete implementations. The advantage is that such abstract solutions fit in unforeseen contexts (new quantum hardware, new programming environments, ...). But patterns represent proven solutions, i.e. by definition they are abstracted from formerly existing concrete solutions. These concrete solutions get forgotten during the act of abstraction.

By retaining (or creating) concrete solutions, making them available, and linking them with those patterns that abstract them, fosters reuse of implementations and speeds up solving problems [9]. Navigating through such enriched pattern languages allows to "harvest" concrete solutions and "glue" them together into an aggregated solution of the overall problem [10].

## 4 Outlook

We intend to grow the proposed pattern language, and make it available in our pattern repository PatternPedia [12]. In parallel, concrete implementations of the patterns in quantum languages like QASM [6] are considered, and these implementations will be linked to the corresponding patterns. Next, we want to evaluate the usefulness of the pattern language based on practical use cases.

**Acknowledgements.** I am very grateful to Johanna Barzen and Michael Falkenthal for the plethora of discussions about pattern languages and their use in different domains.

# References

1. Alexander, Ch., Ishikawa, S., Silverstein, M.: A Pattern Language - Towns Buildings Construction. Oxford University Press, Oxford (1977)
2. Anschuetz, E.R., Olson, J.P., Aspuru-Guzik, A., Cao, Y.: Variational quantum factoring. arXiv:1808.08927 (2018)
3. Biham, E., Brassard, G., Kenigsberg, D., Mor, T.: Quantum computing without entanglement. Theor. Comput. Sci. **320**, 15–33 (2004)
4. Brassard, G., Hoyer, P., Mosca, M., Tapp, A.: Quantum amplitude amplification and estimation. arXiv:quant-ph/0005055v1 (2000)
5. Bruß, D., Macchiavello, C.: Multipartite entanglement in quantum algorithms. Phys. Rev. **83**(5), 052313 (2011)
6. Coles, P.J., et al.: Quantum algorithm implementations for beginners. CoRR abs/1804.03719 (2018)
7. Cortese, J.A., Braje, T.M.: Loading classical data into a quantum computer. arXiv:1803.01958v1 (2018)
8. Dervovic, D., Herbster, M., Mountney, P., Severini, S., Usher, N., Wossnig, L.: Quantum linear systems algorithms: a primer. arXiv:1802.08227v1 (2018)
9. Falkenthal, M., et al.: Leveraging pattern applications via pattern refinement. In: Proceedings of Pursuit of Pattern Languages for Societal Change - The Workshop, Krems (2016)
10. Falkenthal, M., Leymann, F.: Easing pattern application by means of solution languages. In: Proceedings PATTERNS (2017)
11. Farhi, E., Goldstone, J., Gutmann, S.: A quantum approximate optimization algorithm. arXiv:1411.4028 (2014)
12. Fehling, Ch., Barzen, J., Falkenthal, M., Leymann, F.: PatternPedia - collaborative pattern identification and authoring. In: Proceedings of Pursuit of Pattern Languages for Societal Change - The Workshop, Krems (2014)
13. Fehling, Ch., Leymann, F., Retter, R., Schupeck, W., Arbitter, P.: Cloud Computing Patterns. Springer, Vienna (2014). https://doi.org/10.1007/978-3-7091-1568-8
14. Gilyén, A., Arunachalam, S., Wiebe, N.: Optimizing quantum optimization algorithms via faster quantum gradient computation. arXiv:1711.00465v3 (2018)
15. Harrow, A.W., Hassidim, A., Lloyd, S.: Quantum algorithm for solving linear systems of equations. arXiv:0811.3171v3 (2009)
16. Jozsa, R.: An introduction to measurement based quantum computation. Quantum Inf. Proces. **199**, 137–158 (2006)
17. Jozsa, R., Linden, N.: On the role of entanglement in quantum computational speed-up. arXiv:quant-ph/0201143v2 (2002)
18. Kerenidis, I., Prakash, A.: Quantum recommendation systems. arXiv:1603.08675v3 (2016)
19. Lipton, R.J., Regan, K.W.: Quantum Algorithms via Linear Algebra. MIT Press, Cambridge (2014)
20. Mosca, M.: Quantum algorithms. arXiv:0808.0369v1 (2008)
21. Nielson, M.A., Chuang, I.L.: Quantum Computation and Quantum Information (10th Anniversary Edition). Cambridge University Press, Cambridge (2010)
22. Preskill, J.: Quantum Computing in the NISQ era and beyond. Quantum **2**, 79 (2018)
23. Proos, J., Zalka, Ch.: Shor's discrete logarithm quantum algorithm for elliptic curves. arXiv:quant-ph/0301141v2 (2004)

24. Rieffel, E., Polak, W.: Quantum Computing - A Gentle Introduction. MIT Press, Cambridge (2014)
25. Sanders, Y.R., Low, G.H., Scherer, A., Berry, D.W.: Black-box quantum state preparation without arithmetic. arXiv:1807.03206v1 (2018)
26. Thompson, J., Gu, M., Modi, K., Vedral, V.: Quantum computing with black-box subroutines. arXiv:1310.2927v5 (2013)
27. Zdun, U.: Systematic pattern selection using pattern language grammars and design space analysis. Softw.: Pract. Exp. **37**(9), 983–1016 (2007)

# Author Index

Printed in the United States
By Bookmasters